Technology, Culture, Family

Palgrave Macmillan Studies in Family and Intimate Life

Titles include:

Harriet Becher
FAMILY PRACTICES IN SOUTH ASIAN MUSLIM FAMILIES
Parenting in a Multi-Faith Britain

Elisa Rose Birch, Anh T. Le and Paul W. Miller
HOUSEHOLD DIVISIONS OF LABOUR
Teamwork, Gender and Time

Jacqui Gabb
RESEARCHING INTIMACY IN FAMILIES

Peter Jackson (*editor*)
CHANGING FAMILIES, CHANGING FOOD

David Morgan
RETHINKING FAMILY PRACTICES

Eriikka Oinonen
FAMILIES IN CONVERGING EUROPE
A Comparison of Forms, Structures and Ideals

Róisín Ryan-Flood
LESBIAN MOTHERHOOD
Gender, Families and Sexual Citizenship

Elizabeth B. Silva
TECHNOLOGY, CULTURE, FAMILY
Influences on Home Life

Palgrave Macmillan Studies in Family and Intimate Life
Series Standing Order ISBN   **978–0–230–51748–6 hardback**
                                  978–0–230–24924–0 paperback
                                  (*outside North America only*)

You can receive future titles in this series as they are published by placing a standing order. Please contact your bookseller or, in case of difficulty, write to us at the address below with your name and address, the title of the series and the ISBN quoted above.

Customer Services Department, Macmillan Distribution Ltd, Houndmills, Basingstoke, Hampshire RG21 6XS, England

Technology, Culture, Family

Influences on Home Life

Elizabeth B. Silva
Professor of Sociology, The Open University, UK

First published 2010 by
PALGRAVE MACMILLAN

Palgrave Macmillan in the UK is an imprint of Macmillan Publishers Limited, registered in England, company number 785998, of Houndmills, Basingstoke, Hampshire RG21 6XS.

Palgrave Macmillan in the US is a division of St Martin's Press LLC, 175 Fifth Avenue, New York, NY 10010.

Palgrave Macmillan is the global academic imprint of the above companies and has companies and representatives throughout the world.

Palgrave® and Macmillan® are registered trademarks in the United States, the United Kingdom, Europe and other countries.

ISBN 978–0–230–54548–9   hardback

This book is printed on paper suitable for recycling and made from fully managed and sustained forest sources. Logging, pulping and manufacturing processes are expected to conform to the environmental regulations of the country of origin.

A catalogue record for this book is available from the British Library.

A catalog record for this book is available from the Library of Congress.

Printed and bound in the United States of America

To Yara Silva-Tolliday, my daughter, and Natalina Bortolaia Silva, my mother, whose company I enjoyed while doing final revisions of this book by the beach in Ubatuba, Brazil, in December 2009, and for them being in my future and my past, with love

Contents

List of Figures and Tables

Figures

Tables

Series Preface

The remit of the *Palgrave Macmillan Studies in Family and Intimate Life* series is to publish work focusing broadly on the sociological exploration of intimate relationships and family organisation. As editors we think such a series is timely. Expectations, commitments and practices have changed significantly in intimate relationships and family life in recent decades. This is very apparent in patterns of family formation and dissolution, demonstrated by trends in parenting, cohabitation, marriage and divorce. Changes in household living patterns over the last 20 years have also been marked, with more people living alone, adult children living longer in the parental home and more 'non-family' households being formed.

There have also been important shifts in the ways people construct intimate relationships. There are few comfortable certainties about the best ways of being a family man or woman, with once conventional gender roles no longer being widely accepted. The normative connection between sexual relationships and marriage or marriage-like relationships is also less powerful than it once was. Not only is greater sexual experimentation accepted, but it is now accepted at an earlier age. Moreover, heterosexuality is no longer the only mode of sexual relationship given legitimacy. Gay male and lesbian partnerships are now socially and legally endorsed to a degree hardly imaginable in the mid-twentieth century. Increases in lone parent families, the rapid growth of different types of stepfamily, the de-stigmatisation of births outside marriage and the rise in couples 'living-apart-together' (LATs) all provide further examples of the ways that 'being a couple', 'being a parent' and 'being a family' have diversified in recent years.

The fact that change in family life and intimate relationships has been so pervasive has resulted in renewed research interest from sociologists and other scholars. Increasing amounts of public funding have been directed to family research in recent years, in terms of both individual projects and the creation of family research centres of different hues. This research activity has been accompanied by the publication of some very important and influential books exploring different aspects of shifting family experience. The *Palgrave Macmillan Studies in Family and Intimate Life series* hopes to add to this list of influential research-based texts published in English (both new texts and new translations), thereby contributing to existing knowledge and informing current debates. Our main audience consists of academics and advanced students, though it is our intention that the books in the series will be accessible to a more general readership who wish to understand better the changing nature of contemporary family life and personal relationships.

In *Technology, Culture, Family: Influences on Home Life* Elizabeth Silva examines a set of important, yet largely under-explored issues concerning the ways in which developing technologies influence how people's home and family lives are constructed. Sociologists have long recognised that the material conditions of the home are significant in structuring the patterning of domestic life. In particular, changes in housing densities, household amenities, domestic facilities and the like often provided a framework for understanding processes of change in marital and other family relationships, particularly in the decades following the Second World War. In recent family sociology though, the impact of different technological developments on the organisation of family and domestic relationships has been rather neglected, with few scholars directly addressing how technology in general, and particular technological innovations more specifically, influence family practices and the construction of domestic relationships.

Elizabeth Silva is an exception here. In various projects developed over more than 15 years, she has explored the different ways in which technologies – some common, some more radical – used within the home connect with and help shape family organisation and the relationships of domestic life. As series editors, we are delighted to include her book in the *Palgrave Macmillan Studies in Family and Intimate Life* series. In it, she draws on material from her different studies to examine the impact of technological developments on various spheres of domestic and familial activity, including cooking, cleaning, caring work, domestic divisions of labour and sexual activity/expression. Importantly, her analysis goes beyond exploring how different technologies have been used in the respective sets of activities and practices; it also locates the use of technological developments within the relational 'architecture' of the family lives that those involved construct.

Moreover, Silva's book seeks to develop a theoretical understanding of how technology infuses family and domestic relationships. Drawing on the work of a broad range of theorists concerned with different aspects of technology, consumption and family dynamics, she successfully provides an integrated theoretical analysis of how the personal and familial is connected to the material and technological within the home. This, as well as its opening up of questions about the uses made of technologies within family life, marks this book out as particularly novel. It is a welcome addition to the series and will contribute significantly to scholarly understandings of the role of technology in patterning people's everyday home experiences.

Graham Allan, Lynn Jamieson and David Morgan
Series Editors

Acknowledgements

I most gratefully acknowledge the various individuals, anonymised in this book, who offered me their stories, biographical and of everyday lives, with conflictual, happy or painful episodes. Many people also helped me in archives and libraries through funding agencies and refereeing processes. I am unable to name them, but I thank them for their involvement.

Thanks are due to friends and colleagues who made me feel that I was not alone through the many years the projects included in this book have taken to be completed and then to feed into the work assembled here. I was accompanied and nourished by small and big gestures, words and endeavours. Inspiration, cups of tea, a shoulder, a meal, one hour or another of childcare, the cleaning of a room, a sink or clothes, a reading, editing, critique, comment, reference, an invitation to present, discuss, write a piece and so on have all supported my journey doing work on technologies, homes and culture, exploring the domestic in practice and theory.

I want to include in these thanks Lisa Adkins, Raul Antelo, Tony Bennett, Emma Casey, Danièle Chabaud-Rychter, Tony Chapman, Cynthia Cockburn, Marisa Corrêa, Graham Crow, Em Edmondson, Jeannette Edwards, Ros Edwards, Janet Fink, Jacqui Gabb, Jean Gardiner, Ros Gill, Leda Gitahy, Gabriele Griffin, Fiona Harris, Penny Harvey, Clem Herman, Emma Heron, Sue Himmelweit, Helena Hirata, Joanne Hollows, Wendy Hollway, Katrina Honeyman, John Hunt, Sarah Irwin, Tim Jordan, Hannah Knox, Rebecca Leach, Lydia Martens, Cecilia McCallum, Jane McCarthy, Graça Medeiros, Patricia Pinho, Ruth Pitt, Alison Ravetz, Sasha Roseneil, Mike Savage, Sue Scott, Carol Smart, Hillary Silver, Bev Skeggs, Dale Southerton, Pippa Stevens, Lea Velho, Sophie Tayson, Nina Wakeford, Alan Warde, Sophie Watson, Margie Wetherell, Fiona Williams and Kath Woodward.

The series editors for this publication were of enormous encouragement. Thanks to Graham Allan, Lynn Jamieson and David Morgan, together with the commissioning editors, Jill Lake and Philippa Grand, and the editorial assistance of Olivia Middleton, as well as the learned copy-editors, in particular Jon Lloyd, who helped to sort out my text. Special thanks due in this regard and in others of a personal nature are given to Steven Tolliday because of the lengthy duration of this work.

Introduction

Why does the profound technological transformation of the late twentieth and early twenty-first centuries appear to have done so little to effectively change patterns of care and ways of living in the home? Why is it that long-lasting social divisions are continually 'reconfigured' in different contexts amidst a vast array of choices for living differently? Do sexism and power continually shape the meanings and patterns of use of even the most radical technologies? If change exists both in technological developments and in ways of living personal lives, why do these seem to occur in parallel but remain relatively disconnected from one another? And why is it that the world of technology is usually seen to 'impact' upon personal ways of relating and of doing things rather than vice versa?

This book is about understanding personal, relational and material matters in everyday life in the context of broader and long-standing social problems and issues. I use stories that I have gathered during many years of research into the ways people live in their homes in the UK at present to reflect on the connections between everyday practices in the reproduction of our bodies and our relations with those we live with, and the technological patterns and practices of a world driven by forces that go far beyond any individual or small group.

I deal with broad sociological questions from a perspective focused on ordinary everyday practices: how we nurture, clean, care and choose how to do things. In what sorts of material environments do we do things like cooking, cleaning, consuming and caring? What sorts of tools and technologies do we use to carry out these activities, when we do them and why do we do them in these ways? How are the technologies embedded not only in our practices but also in our ways of relating? What are the practical and moral implications of the ways we do things for the relations we form with each other and the social worlds we build?

Both theoretically and in the real world, unravelling the connections between grand social life issues and those of mundane everydayness keeps the social scientist on her or his toes: they are complex, often surprising and unpredictable. Disrespecting disciplinary boundaries, I draw from various traditions of thought. My mode of questioning and reasoning combines three voice registers. I use social theory and the reasoning of academic texts, examining authoritative voices which I sometimes endorse and at other times dispute. As an author I mostly use this voice. However, my own voice also appears as a second register, in a biographical mode, in a colloquial style of reasoning and relating with participants in the project. A third register consists of the research material generated in the investigation, which is of two kinds: (1) archival material of manufacturing innovation and developments in technology and media texts; and (2) the narrative of the research participants in the ethnographic study. While the first is registered in a more formal way as evidence, I use the second in quotes and in registers of silence, or in the narrative I construct about practices observed. The assembly of these voices is aimed as an explicit gesture towards a way of doing social sciences that seeks to engage the (unknown) reader, while giving access to the lived experiences of those studied – and of the one studying – which inform the analysis. The more usual formal and distant voice stylised in academic discourse is of great relevance here for the definition of concepts, the following of procedures and the armoury of tools developed to attain scientific knowledge about social worlds. I hope that the weaving of these voices is attractive or that readers may choose which ones to engage with, using their own criteria about what and how to know. I believe that while depth and complexity are achieved by following the whole text, readers can grasp the key arguments by 'hearing' more closely the voice of their preference.

I put some theories to work, testing and illustrating their shared and diverging directions. I developed a 'conversation' between the empirical material of UK contemporary home life and the work of influential French academics – Pierre Bourdieu and Bruno Latour – and some of their critics. These authors have been frequently taken up in recent Anglo-American social science debates, but seldom in relation to each other or in reference to personal and relational matters in home contexts. These are not areas of interest to Bourdieu and Latour. Yet, their ideas are productive in providing a framework for exploring the connections between the material (both the material basis of existence and in relation to the form of technology, for instance) and the social, and for directing attention to how ordinary practices connect and constitute the social world. I engage these theories with a firm claim that home and family life are areas of intrinsic significance for the workings of the world which must be taken account of in mainstream sociological narratives.

On the basis of the investigation in this book, I deal primarily with six lines of argument: some I contest, some I build upon, and others I present as a more innovative way of thinking about technology, culture and family:

1. A long-standing argument in relation to household technologies is the paradox that personal, relational home lives seem to be so little affected by technological developments, leading to the consequent reinforcement of gender roles. What features of these dynamics continue to be found in a world where women increasingly work for pay outside the home? Where and how does social change in technological innovation for homes emerge?

2. The argument regarding the persistent reproduction of social divisions appears to be hard to sustain in the face of increasing change, evidenced by diverse ways of living. How are these tensions embodied in contemporary technological home living? How can change in this area be captured in research? What sorts of change are captured from this diversity? How do changing social divisions relate to the materiality of the social world?

3. Leading ideas about the operations of power in the social world, such as the view of power as capacity and effectiveness, as seen in the work of Latour, or power as domination, as seen in the work of Bourdieu, cannot capture the nuance of everyday negotiations involved in living togetherness (in all its varieties). Newer theories are needed to refine the use of these ideas so that they can inform theories about the operations of the social.

4. Family life, as much as personal lifestyles, is built around and upon active engagement with different technologies which are in some respects gendered or classed, and which depend on access to material resources. Technologies are resources in themselves and as such they also construct social positions and social divisions and are deeply implicated in social life.

5. Methods of investigation are formative in the research findings and the theoretical ideas developed. Sociological claims that technologies 'impact' upon family life exemplify this, as they derive from a vision of cohesion, homogeneity and consensual dynamics within households where power appears to be unidimensional and somewhat uncontested.

Through ethnographic enquiries it is possible to account for subtle and messy relationalities.

6. The task of unravelling the interactions between material, cultural and social conditions and those of personal and family lives is best accomplished through the use of evidence and documentation to show how things and relations work.

My engagement with these arguments generates three main areas of investigation: (1) *the significance of ordinary home experiences* in relation to the patterning of social life like gender and social divisions; (2) *the relation between the material and the social* (understanding the material as technological as well as economic and environmental matters), and (3) *the resources involved in relationality* like social positions, values and objects (especially technologies). The Bourdieusian notion of practice, which explains the methods of how to proceed in social life, is central to the development of my arguments and is interwoven into the concepts of 'everyday cultures', the 'subjective' and the 'material', which anchor the analysis. Practice is implicated in activities of making homes, taking care of the self and others, in cooking, cleaning and taking decisions about what to do, such as the implementation of routines and the allocation of time. 'Practice' is also implicated in ways of seeing the world. The world is complex, but we make sense of it through regularities, through practices guided by the habitus (those culturally learned schemes of perception and action) which are prisms of reality. To capture 'practice' empirically, I use Latour's approach, in particular regarding a view of the social world embedded in technologies – and in our relations with them.

The three main areas of investigation outlined above and the related approaches to knowing about them are pursued in each chapter in relation to the various issues and literatures involved. In each chapter the relevant arguments are outlined and then put into conversation with the empirical findings from archival research or interviews and observational studies.

In Chapter 1, I set out the ways that I employ issues of materiality and the subjective, of everyday cultures and family practices. Drawing from an academic feminist perspective on household technologies, I discuss the insights that the works of Bourdieu and Latour bring to developing new theoretical ideas in this field. I acknowledge the tensions in combining their visions of the social, and I both qualify and depart from their propositions. Bourdieu's vision of the relationality of the social centred on position in social space is important. Yet, as I argue, the social in Bourdieu has two major shortcomings. One is his conceptualisation of gender narrowly derived from the sexed body. The other relates to his narrow vision of the 'normal' when reflecting on family life. Latour's focus on the human and 'non-human' elements in making up the social provides me with a way of thinking about connections between individuals and machines in terms of scripts for action and descriptions. The key limitation in the work of Latour centres on his privileged focus on a masculine and powerful world, which presents a very incomplete view of the realities of social life, particularly when it is applied to family and everyday concerns. Also, Latour re-inscribes his vision of the socio-cultural onto the non-human, a move I contest. While key concepts and frameworks provided by these two theoreticians are useful, I show that their ideas need to be modified when they are used to focus on families.

Chapter 2 considers the materiality of homes and the identity processes of individuals in relationality, and addresses claims about epochal social changes related to technological transformations. I explore theories, point to their failings regarding the understanding of home life within cultural processes, and suggest that a combination of statistics, descriptions and case studies offers a productive way of accounting for refined processes of change where individual adjustments are important for the collective socio-cultural and material patterns of contemporary life. This proposal also requires a revision of the notion of practice, moving it beyond Bourdieu's approach to accommodate a place for the interplay of 'localized dynamics' (cf. Thévenot, 2001). The chapter provides an account of developments in the infrastructure of housing and public services, the production of technologies used in the home, their mode of consumption, and their connections with changing family forms. Linking material life with relations between people in the context of family life, this is also an important background chapter for the subsequent discussion of relationality within material contexts. Stories of the household lives of Londoners – potentially cosmopolitan and 'on the edge' lifestyles – provide interesting cases in which to explore the relationships between material possessions and position in social space, including the lifestyle and identity of the families and individuals concerned. I show fixed and mobile gendered and sexual identities operating in these contexts where individual practices are coordinated with 'localized dynamics' and are still patterned in wider socio-cultural and material contexts.

The issue of social and technological change is further pursued in Chapter 3, where I outline, and take issue with, the theoretical concerns and findings of 'classic' studies on household technologies, mostly centred on the use of women's time in the home, and more recent research on gendered uses of time, accounting for cultural diversity and contemporary changes in the ways that lives are lived in the home. I analyse in detail the routine patterns of the families in my study, showing that for many people it is difficult to trace the boundaries between work and non-work time, time for care and for leisure, as well as the uses of technologies for education, work, pleasure, relationships and so on. While this blurring is more prevalent among women, it includes individuals who work from home full time or part time, who look after children, who work in the media and cultural sectors, or who have more flexible working patterns. This reflects changes in practices of masculinity and femininity in the home, which are not circumscribed to family living. Market conditions, commercial arrangements and indeed technological design and innovation are also implicated in these processes.

The ways in which household technologies are constructed in relation to certain dispositions and practices related to normative expectations of gendered everyday life in the home are explored in Chapter 4 in relation to cooking. The technologies examined are cookers, fridges, freezers and microwave ovens. Cooking in the family involves various expertises. Practices are linked to a 'technological nexus' involving the use of restaurants, school meals and hours of employment outside the home, as well as utensils, washing-up demands and so on. Convenience food illustrates the ways in which the 'technological nexus' operates, closely linked to levels of wealth and lifestyles. Cooking is one of the most traditional areas affected by technological innovation in the home and is one of the activities that has undergone the most change, due to recent technological developments associated with new family practices. Based on archival material of manufacturing processes and media texts, I discuss technological change with a focus on arguably the two major innovations in cooking in the twentieth century: the advent of the thermostat oven control and the microwave oven. I also explore current feeding, eating and cooking activities in the home to show the connections between 'localized dynamics', involving the resources and experiences of cooking, and wider patterns.

In Chapter 5, practices of cleaning are investigated regarding the change over time in the instruments available for laundering and dishwashing. The focus is both on technological developments and cleaning practices in the context of care of the self and others. The idea of a homology between social position and consumption practices is explored on the basis of two contrasting ethnographic stories. In the comparison, five themes stand out challenging established theories: (1) ways in which practices to save time or use time relate to personal lifestyles and the requirement of resources in how to live; (2) the relative roles of technological scripts in personal choices and localised dynamics in sustaining or changing gendered roles; (3) the choices of individuals to ignore or use the capabilities of machines; (4) the uses of technologies to create either order or disorder; and (5) the efficient function of technologies competing with other ways of inserting them into home life and giving them value. Technologies in the home are shown to be crucial 'relational resources' which are, however, patterned according to hierarchies of position in social space.

Centred on a discussion of consuming and caring, Chapter 6 addresses questions of which resources are felt to be necessary for everyday domestic life and where these can be drawn from. Time and money are core resources for caring in the everyday, but so are personal connections, emotional states and abilities. Two stories of the acquisition and use of kitchen technologies are discussed to illustrate these issues. In both, economic resources are scarce and access to technologies is achieved by different means. One case illustrates a strategy of emotional investment in a choice of lifestyle, signalling upward mobility. The other refers to unfulfilled desire in a context of great deprivation, where the availability and choices of resources mirror a personal lack of emotional, social and economic resources. Here again, inequalities in social position are powerfully linked with the material resources found in the home. Resources for consuming and caring are thus associated with senses of a valued self (and related dispositions) as well as with assets derived from emotional resilience, social connections and technical expertise.

Domestic dilemmas, the core issue in Chapter 7, are used to explore the resources that inform moral values implicated in practices of everyday living. What practices are emphasised and how are different practices justified? In what ways do choices relate to the social position of individuals? The continuous need to change according to life circumstances, the difficulties of changing other people's practices and the importance of connections with children are the key issues considered. Dilemmas are explored on the basis of three vignettes (from the ethnographic study) related to judgements of actions by individuals using computers, video-recording equipment and their involvement in general domestic activities in the home. The expressed judgements convey important aspects of self-projection and identification that communicate nuances of social positions in the ethical dilemmas and valuations of how to live interdependently.

Chapter 8 focuses on sexual lives in our technologically-drenched everyday culture. Sexual practices and their involvement with technologies tend to be naturalised in the everyday. Ordinary sexual practices are made invisible within a politics of pleasure that gives primacy to danger and transgression. While taking issue with this invisibility, which is evident in contemporary life and in academic theory, I investigate connections between technological and sexual stories in my ethnographic study. New technologies for reproduction and for communication are discussed in relation to the stories of research participants. Views about the impact of the 'information age' on personal and relational lives are challenged using this empirical material. My emphasis is on the importance of building on an approach that, by being sensitive to personal matters of a relational kind and by focusing on material and culturally elusive ties, can account for meaningful, inconsistent and paradoxical individual stories constituting our collective history.

Following David Morgan (1996: 10–11) I refer sometimes to 'the family', to the 'household', to 'domestic life' or 'home life', or to 'home'. I recognise that these are not synonyms and my critical use of these words aims to contribute to a sense of fluidity and flux in family studies, using the terms to refer to sets of practices linked to the notion that family is an important influence on the ways in which individuals understand and structure their lives. 'Family' is discussed as a practice, an interdependent relationality and a setting. The narratives of family practices in this book are captured by the observation and narratives of individuals and documents of various sorts, including interviews, video recordings, archival materials and secondary data and analyses. These narratives, centred on everyday practices, are mostly reached through stories that people tell about themselves. While these individual stories draw on public stories as a mechanism by which sense is made of personal life, they also indicate something of a direction of social change (Jamieson, 1998: 159). The stories I wish to unravel here show how culture and technology are resources for personal relationships in practising family as well as resources for wider social, economic and political purposes.

The material I draw from consists of several research projects that have used a variety of research designs. The first, starting in 1996, comprises a study of the innovation patterns of household technologies and associated changes in gender relations since the beginning of the twentieth century. This was based on archival material and interviews with key people engaged with the design, manufacture and marketing of selected technologies. The second, begun in 1998, is an extensive and in-depth study of home life in the context of the use of technologies for housework, education, work and leisure. This ethnographic study includes 24 families, 16 in the original study, which was expanded with a further eight to take account of a cross-sectional representation of contemporary family types. Both the 1996 and 1998 projects were funded by the Economic and Social Research Council (ESRC). In the methodological appendix, details of the sample, fieldwork processes and the profile of research participants can be found.

Summing up the ways in which contemporary home life is drenched in relations with the technological, I revisit in the conclusions the central contribution of Bourdieu and Latour to the discussion of technologies as resources for living, defined in cultural terms. I examine their conceptions of how the world works, in particular regarding the social divisions informing gender practices, the notion of power and the role of the habitus, and reflect about their fit with contemporary practices of families and selves, as found in my empirical investigation. My focus on family life creates a particular take on the theorising of the operations of culture and processes of social change. Following the three central areas and questions pursued in this book, ideas are outlined in reference to: (1) the interplay of ordinary living and social patterns relating to immediate occurrences and to long-term processes of social change: how do broad patterns and minute everyday practices connect? How does what happens here and now relate to past and future processes?; (2) the relations between the material and the social when 'things' appear as resources: how do technologies operate within social divisions and how do they change existing patterns of relationality?; (3) the resources used in relationality regarding both concrete practices of lived experience and practices of knowledge construction: how can we account for agency (both of individuals and of that embedded in objects), creativity and contextual adaptation that is evident in the practices of individuals and families? What is the significance of emotions and ethical dispositions in the choices made of how to act? How are change and permanence disputed both in theory and in the everyday lives of individuals and families?

1
Relations

For many years I have been thinking, researching and writing about cultural and material life, mainly framed around issues related to technology in the home and meaningful close relationships between individuals, which I describe as being about 'family' (of all types). I have an image of this book as a sort of map of the various 'sites' I have visited during this intellectual journey, much of it tied up with my personal biography. It has been about going forward but also about turning back, about collecting words and thoughts between me and my academic interlocutors and about my ethnographic engagement with people in different cultural contexts over many years. The book narrates episodes from the lives of individuals and describes the ways in which subjectivities and everyday events are attached to the technological materiality of the contemporary world. My attention in this book is captured by the relations between the phenomena of a technologically cultured world and the singularity of lives and close relationships. Central to this is the question of understanding social change from the perspective of the ordinary, which is often overlooked.

The workings of the three concepts of 'everyday cultures', the 'subjective' and the 'material' are outlined in this chapter. Tying them together is the central notion of practice. In Pierre Bourdieu (1992a), the 'logic of practice' provides a means of understanding patterns which are rule-like, but are not prescriptive or deterministic. Practice is about personal methods of knowing how to proceed in social life. These 'methods' are reflexive and inter-subjective. As Margaret Wetherell (2006) puts it, method is a most pertinent term in relation to practices since, while being open-ended, methods are organised, logical but flexible, potentially self-conscious but frequently unconscious and habitual. Practice also has a time dimension, being conditioned by the past, yet indicating expectations of current or future accomplishments. Because practices allow for the exploration of traceable patterns over time (they are somewhat predictable in their continuity), they indicate change. Importantly for my work with the personal, the cultural and the social, practice is a concept that bridges and simultaneously enables the investigation of collective patterns from individual narratives.

I engage with the concepts of materiality, subjectivity and everyday cultures from the writings of feminist academics in various fields and more broadly from sociologists, analysts of culture (from literature, anthropology, history and cultural studies), feminist academics and philosophers of science. The works of Pierre Bourdieu and Bruno Latour inform the theoretical discussion about the intermingling of the object (things and technology) and the subject (human and personal or individual) in the social world and in the everyday, anchoring the analysis. I find that the differences in the positions of these authors in relation to the social and to the means of its investigation are productive in furthering the analysis of connections between the three concepts, and their individual insights guide my investigations.

I introduce here the theories of Bourdieu and Latour as they relate to the issues in the book. The connections between materiality and subjectivity are explored in the next section. How do objects feature as social agents being active in culture and implicated in politics? How does this activity become naturalised – and invisible – but remain central to subjective experiences? In the following section I consider everyday cultures and 'family'. How can the researcher capture ordinary practices and how are social patterns discerned from messy everyday living? In the last section I outline the research practices undertaken in this book with fragments of my own personal trajectories in family and as an academic seeking to understand personal lives and relationships within technological cultures.

Materiality and subjectivity

In our relationships and subjectivities, the ways of doing things and the objects we live with are intimately related, though we are often unaware of the workings of technologies in our lives. When objects are created, or when they encounter new users, break down or provoke an accident, they focus attention on how they feature in people's lives. But when our experiences with objects are taken for granted, objects recede into the background and it is more difficult to bring them back into the light. They become part of a 'seamless web' (Hughes 1983) in which lightbulbs, microchips, cookers, televisions or computers become invisible. It is thus perhaps understandable, though clearly a great neglect, that social scientists have mostly discussed a predominantly object-less social world, even though our everyday routines are accompanied by passionate affairs, contiguous and continuous intimacy with objects.

The relationship between things and people has been central to the work of both Bourdieu and Latour, who have, in different ways, sought to break down conventional categories and scientific understandings about the workings of the social world. They deal with the idea of the thing/object in very different ways, which I will outline below. Their approaches share the desire to make visible what is hard to pin down: those relationships that concern complex, messy, hidden and heterogeneous realities. They also share a concern with how particular practices of social scientists shape the social as part of their explorations of the ways in which 'intellectual technologies' (documents, writings, charts, files, paper clips, maps and organisational devices) are used. They work predominantly in different 'fields': Bourdieu on the production and appreciation of aesthetics (the consumption of art, literature and taste as marks of social distinction) and Latour on the production of science and technology (engineering and the laboratory). Their fields differ from my interest in technology in everyday home life, but I draw from their work with the aim of developing some shared concerns. While showing connections between their approaches – and with mine – I note what is missing from them.

Bourdieu treats objects as social agents in the same way as he treats the world of art, music or literature. They are active parts of culture, a dimension of social life with huge political implications, and are never innocent. In Bourdieu (1977), subjects (humans) are as much the product of objects as objects are the product of subjects and the subjective. These connections are marked by entrenched processes of reassertion of class divisions when material goods – for example, objects to decorate the home – objectify the values and moral order of the social field. He uses statistics and documents to show the varied and subtle ways in which access to cultural objects is the privilege of the cultivated class, of their values and morals, including the maintenance of their privileged position. In his largest empirical study, *Distinction*, Bourdieu (1984) analyses the significant differences of engagement with diverse objects according to the social class of the individual. This has implications for the analysis of technology innovation and consumption in households.

Bourdieu argues that individuals of different social locations are socialised differently, reproducing the values of their social position of origin. This socialisation gives individuals a sense of what is natural and comfortable (he calls this *habitus*). These early experiences shape the amount and forms of tradeable assets and resources (which he refers to as *capital*) that individuals inherit. Inheritances from birth and other people, acquired through processes of socialisation, are drawn upon as individuals deal with various institutional arrangements (he calls these *fields*) in the social world. Capitals of various kinds – social (from connections), economic (from financial assets) and cultural (from education and early socialisation) – reinforce each other in this process. For example, the cultural training in the home is transferred to other institutions where the standing of the middle classes, which is dominant in the institutional realm, affirms the standards of the privileged, asserting a pattern of domination and inequality. In this process the transmission of privilege is 'misrecognised', as individuals tend to see their society's social arrangements as natural and legitimate. The domination of the powerful is thus reinforced in an invisible way in the relationships between individuals and between individuals and objects.

In Bourdieu's framework, by means of the principles of social classification, the attachments of individuals to the material world – what people choose to have and live with – both derive from, and reassert, their social position. By tracing the history of individual or collective taste, it is possible to show the limits to the notion of the natural creation of preferences and to reveal and make visible the myth of innate taste. The objects with which we surround ourselves thus appear as a product of our own place in the social, indicating homology between taste and social position. Bourdieu reads a close connection, revealed by tastes in furniture, décor, food, art, literature and so on, between social classification and individual classification, which builds social continuity and cohesion.

The implications of Bourdieu's framework for understanding the processes of production and consumption of household technologies are significant in relation both to the gendering codes built into innovation processes of design and production and to the class significance of the consumption of different artefacts. As technologies are structured by human practice to structure human practice, they embody in their physical form particular dispositions (ways of doing things by particular kinds of people). If Bourdieu's idea of a homology between social position and individual engagement with objects is plausible, it is relevant to enquire how far this homology goes. Apart from class position, what else is important regarding the relationships between individuals and objects? Expanding this further, it is relevant to discuss the significance of age, ethnicity and sexuality in the complex patterns of socio-cultural divisions of social life. These divisions, more often than not, also compound the contemporary habitus in more significant ways than class position (see Bennett *et al.*, 2009). My aim in exploring connections between individual life and material contexts in the social world involves verifying the applicability but also the limits to Bourdieu's homology assumption, in particular as far as gendered patterns are concerned (Silva, 2006).

Andrew Sayer's (2005) work is important in this regard. He proposes an expanding Bourdieusian concept of the habitus, in which the ethical dimension of the self and of relationships – how people do, and should, treat one another – is often more important than the functional and aesthetic dimensions of the social field. The valuations of goods and their holders are thus linked together in the struggles over ways of living. Positions refer not simply to possessing this or that but relate to the moral concerns involved in living in one way or another, where relations of dependence upon others take prominence. Sayer places evaluation and choice at the centre of the subjective experience of social positioning, instead of social positioning being an immediate direct consequence of one's origin, as Bourdieu understands it. I embrace Sayer's assertion that while objects are valued for their properties, valuations relate to both objective and subjective aspects, and will highlight these in the case studies that follow.

But how is it possible to capture empirically the active engagements of individuals in this process of choosing how to live, and what to live with, in particular social fields, and specifically in 'family life'? The framework developed by Latour is a useful tool in this regard.

One of the questions explored by Latour is the delegation of human action to the technical or, in the Latourian vocabulary, the 'non-human'. One of his paradigmatic examples is the automatic door opener, through which he demonstrates how our bodies are unnoticeably disciplined when we practice skills and capacities that become habitual, such as pushing doors to open and then relying on the automatic closing system of the 'door groom' (Latour, 1988b and 1992). These little interventions increase or decrease body capacity and skills, rendering many old gestures and expectations automatic and invisible. More versatile, cheaper, more constant and disciplined than a porter, the technology of the door opener saves the investment of disciplining a person into a porter. But the door opener also discriminates against the weaker and the frail: children, the elderly and people carrying packages have difficulty negotiating the standard use prescribed by the innovator:[1] the spring mechanism is too strong and the time set for opening and closing is a standard one. Showing the moral prescriptions incorporated in artefacts (for example, who is to be discriminated against or favoured), Latour claims that technologies are mediators which can change social action. This is not only because they imprint and demand specific behaviour, but also because, in the terms used by Michel de Certeau (1984), the discipline envisaged through the operation of the object can be subverted by the use of tactics, changing its use and its story.

The concern about technology and social change aligns with a large debate in studies of science and technology. Historian Langdon Winner (1988) addressed this issue, asking in the title of an early article 'Do Artefacts Have Politics?'. The answer is that artefacts are embedded in political intentions, the contours of the desired social order tending to be strongly fixed into material equipment. The point is emphasised by Donald Mackenzie and Judy Wajcman (1985/1999), who stress that technologies can be conceived, consciously or unconsciously, to open certain social options or to close certain others. Latourian sociology, also identified as actor-network theory (ANT) and developed by various contributors (including Madeleine Akrich, Michel Callon and John Law), explores these concerns, leaving out the preoccupation with power differences which is prominent in Winner's formulation. ANT claims that actions are written into things which in turn become resources for moral delegation or further inscriptions. Through the technical, social and political, rules are dictated, which structure the social in durable and invisible ways.

The robustness and power of the technological has been a central question examined by Latour (2005). Although many of the properties of technological apparatuses are linked to models of power and authority prevailing in social life, Latour assumes that things do not have power in themselves but assume power in the way that they are associated and distributed (Joerges, 1999). Things are thus resources for forming the social. Certain technologies are more compatible with certain social relations than with others and, in adopting them, we also adopt most of the economic, social and political orders they carry with them. But the main political character of technology is in how its appearance hides its human, social and political fashioning behind the social process of naturalisation and neutralisation of the artefacts (Akrich, 1987).

The question of the reproduction of the social, though common to both Bourdieusian and Latourian frameworks, is understood differently by each of them. Bourdieu emphasises taste ingrained from birth and early socialisation as products of habitus, as forms of cultural capital or as lasting means of reproducing class-based hierarchies.

Latourian ANT poses different questions, where individuals appear engaged in activities that make and remake their competencies and capacities. The often relatively passive and unknowing individual of Bourdieu, behaving basically according to the constraints of her or his habitus, differs from the active and competent individual of ANT. But, as Cynthia Cockburn (1992: 39), following Susan Leigh Star (1991), remarks, the ANT agents are basically active powerful men, major projects and important organisations, while the 'excluded' (children, women, the elderly, disabled or weak – for instance, see Latour, 1988b, 1996, 1999) are dismissed. Informed by Michel Foucault, in ANT power is represented as capacity and effectiveness, but power as domination remains invisible. While Latour affirms that power should be treated as 'a consequence rather than a cause of action' (Latour, 1986: 264), Bourdieu stresses the centrality of social hierarchies and of class habitus – acting as embodied social history – in social action. There are thus powerful issues in tension in these two approaches.

I retain from Latourian thinking the idea of the agency of humans and objects, and of the inscription of social action in the technologies. In this idea the user is integral to the process of technological development. However, as noted above, feminist research has remarked that while ANT perceives that artefacts embody the relations that went into their making, the inevitable gendering and empowering (including the power of class hierarchies) of this process have not been given proper attention (Cockburn and Ormrod, 1993).

In the feminist informed literature of science and technology studies, there are many suggestions of the manufacturing of gender and of power relations which counter the shortcomings of ANT in these aspects (see Chabaud-Rychter and Gardey (2002) for a review). Things/objects contribute to the social (and the gender) order. This includes the construction of differences, the interventions to modify the body of individuals and to change their subjectivity, the fact that things can be gendered and attributed to particular sorts of persons, and so on. For example, Cockburn (1985) observes how the gendered body is socially produced through social practices and how minute differences, from the point of view of physical force and technical capacities, are developed as advantages for men and disadvantages for women, and are then multiplied by their differential access to technologies. The appropriation of muscles, competences, instruments and machines by men are means by which women come to be defined as women. In the world of work, argues Cockburn, the deskilling of women achieved through technological innovation is part of the power game of masculine domination. Confirming this logic, I have noted, as part of the same power game, that in operations of technologies in the home, the skills of women are made invisible and remain unacknowledged (Silva, 2000a), even when these skills are in practice essential. I shall elaborate further about this later in the book, in particular in Chapters 4 and 5.

From Bourdieusian thinking I retain the concern with the pervading effect of entrenched social divisions of class informing concrete practices and processes of evaluation. However, within a broader feminist critique of Bourdieu's engagement with the constitution and effects of gender in the social field, I stress that his framework needs to be adapted in order to understand contemporary family life, in particular his idea of a 'normal' family within the processes of production of social hierarchies (Silva, 2005). While I do not want to fully rehearse here the argument concerning the applicability of Bourdieu to the analysis of contemporary family life, it is relevant to bring in Lois McNay's (1999 and 2000) engagements with Bourdieu to highlight embedded and embodied aspects of gendered and classed identity as unconscious and pre-reflexive practices, which are played out in everyday life. These practices explain the entrenched force of conventional arrangements of gender in families, despite the objective shifts of the social position of women derived from their increased participation in the labour market. Practice is a method of knowing how to proceed in social life. I explore ordinary practices in everyday life using Bourdieu's work towards the goal of developing a framework for understanding contemporary family life and for reflecting about the importance of varied cultural resources of social position across different kinds of socio-material relationships.

Unveiling how particular gendered and individual engagements (including generation, class, sexuality and ethnicity) with the material world convey distinctions and are played out in the context of hierarchical and patterned cultural and social life, I aim to connect, in a Latourian sense, the operations of micro-units (families and individuals) and the operations of the wider culture.

Everyday cultures and 'the family'

Puzzling over what human beings do or do not do in ordinary situations reveals a preoccupation with everyday life. Asking to find familiar behaviour strange, or not strange, is a particularly common heuristic of ethnographic studies. This book describes the ways in which everyday life is attached to events of personal and wider orders. How can we recognise in the objects that are familiar to us and part of our everyday life – for example, a microwave oven – settled practices, knowledge and technical templates typical of a particular social order? How can the individual use and maintenance of an object someone possesses – a dishwasher or a computer – connect to one's gender and bodily experiences, or to one's view of her or his role in a family? The notion that things are produced brings about the idea of a fabrication, a construction, signalled by the mingling of the technical and the social. If technologies are made socially and are consumed in particular social circumstances, it is possible to trace connections between them and people in explorations of everyday life. But dealing with different case studies brings into question the roles of parts, fragments and the boundaries or limits of 'a totality'. I do not seek to make a picture of a totality but to consider particular ways of engaging with the world. As Veena Das (2007: 7) remarks, when thinking of agency in terms of forms of engagement, rather than as a descent into the ordinary, our theoretical impulse is often to think in terms of escaping the ordinary. Here, I counter the suspicion of the ordinary by paying very close attention to the most ordinary of objects and events in everyday family life.

To explore the potential of the approaches of Bourdieu and Latour to the study of technology, culture and family, I turn to consider how their frameworks provide insights into the analysis of the position of the family in the social. As Bourdieu and Latour introduce innovative ways of discussing the relationship between things and people, breaking down conventional categories, they raise questions of how a concept of 'mixed sociability' can be used in practical studies of technology and family life. But they are not concerned with families or homes in their explorations. I on the other hand am and, while building on their contributions, my questions are different. How does one develop knowledge about the internal life of families? It is often difficult to engage in talk about personal subjects, but studying technologies in family life provided me with a device for entry into broader talk by discussing objects and concrete activities as social practices.

The concern about how things and humans 'talk' is central to my exploration of technology in the family.[2] Humans speak. But the disjuncture between certain kinds of lives and language, or the fit of experience with certain contexts of talk or vocabulary, often requires translation. Moreover, linguistic incongruence is found in matters of gender, class, ethnicity, nationality and in the very process of listening to talk (DeVault, 1990). Humans speak in ways limited to hierarchies of social position, by loud and muted sounds, through bodily gestures and emotions. The complexity of human communication also means that talking plus listening does not directly result in understanding. The problem is increased for researchers because the invisible structures that organise talk in research situations often derive from different social spaces, the academic space being notably present in the capturing of talk (Silva and Wright, 2005). Objects actively engage with humans and social worlds and we could say that they 'speak' when they are created, that their 'talk' is often disclosed in studies of innovations, and that they also 'speak' when not taken for granted and brought out of the 'seamless web' of which they often become a part.

Social research like this does not only face the problem of understanding how humans and objects 'speak'. Research itself also 'talks', producing the social in a wide range of locations (Law and Urry, 2004; Law, 2004). Nothing 'speaks for itself', as Bourdieu says (1999: 621), and Latour (1999: 139) argues in relation to the practice of doing science that 'we cannot wait for light to fall on things'. For some time it has been acknowledged that the social sciences work upon, and within, the social world, helping to make and to remake it (Osborne and Rose, 1999). Of course, it does so initially through the hands and minds of social scientists, and its work is then filtered through government agencies, the media and everyone else, becoming part of the social (Latour, 2005). Equally, the minds of social scientists involved in this social making are themselves part of the social.

Bourdieu (1988) claims that theory is performative since classifications make effective the perspective of the classifier, or of the researcher who classifies to bring about intelligible patterns and relationships. This means that the process of knowing demands knowledge of the conditions of the creation of knowledge (Bourdieu, 1992a, 2004). Theoretical intentions and research practices that are often kept separate need to be reconciled by means of a critical approach to the theory where the parameters of the conditions of theoretical production are made visible. How? In the constitution of a particular object of observation and analysis, the process of knowing is practised and delineated within the limits of a particular vision in a particular time and space. This involves both external and internal limits, which relate in different ways to the habitus and dispositions of those involved in the 'story'. The habitus, as incorporated history, provides practices with relative autonomy in relation to the external pressures of the immediate present. Dispositions, though basically referring to internal processes and subjectivity, are objectified in durable agents (like institutions) or things (like technologies) to satisfy the demands of a specific field (home life or family, for example).

There are disagreements between Bourdieu and Latour over the ways in which scientific accounts are created. Bourdieu greatly impoverishes Latour's work in the interpretation he offers of it. He reduces

Latour's work to the demolition of the distinction between human agents (or forces) and non-human agents, arguing that the principle of actions guiding the production of science should be sought in positions and dispositions, not in description of alliances and struggles for symbolic credit (Bourdieu, 2004: 26–9). On the other hand, Bourdieu's attempt to explain the social is disdained by Latour, who claims that 'the social has never explained anything; the social has to be explained instead' (Latour, 2005: 97). For Latour, the social has no role in constituting reality, as it is generated in the very process of discourse, for example, by saying that 'an A is related to a B' (2005: 103), thus producing the connection.

For Latour, the apprehension of reality is made by 'following the actor' and thus drawing networks of connections. He seeks to *describe* the operations of technologies and individuals and the ways in which they connect. The tendency is to put everything on the same *plain* level and notice the way they work. What counts is what is said, as a plain fact, and the way of saying it. Bourdieu engages with a different empirical project seeking to *analyse* how things work. What is said is not taken at face value. He searches for *deep* underlying meanings, uses, functions and origins of what is said and done, seeking to unravel how the vision of the social relates to the division of the social world. While for Latour, echoing Foucault, all relations, including those between humans and objects, are relations of power – expressed in capacity and effectiveness – and reality is a social construction (a textual creation with no objective life), Bourdieu contends that power – expressed in hierarchy and domination – is specific to context, where realities of structures and dispositions prevail.

Regarding the different approaches to seeing social life, I align more closely with Bourdieu, particularly in seeking to explore the underlying causes for social divisions. On top of the commonalities already outlined between the approaches of these two authors, another link concerns the roles they attribute to dimensions of agents: the macro-micro in Latour and the micro-units and the structure (or the field) in Bourdieu. Ultimately, they both recognise that a series of interlocking micro-cosms exist which, speaking in Bourdieusian language, have relational positions and associated effects of positioning. Latour's proposition is helpful in seeking to associate these by tracing connections through des-criptions. Bourdieu's formulation usefully addresses the concern with the political via games of positioning, power and social struggles. His notion of hierarchical power, and its connection to domination and symbolic violence, counters the flatness of ANT's view of social relations, allowing the associations between humans, and between humans and objects, to be traced in a more stratified and differentiated view of social reality, restoring human primacy in relation to non-humans. After all, even though things make humans act (that is, do things), humans act with things in a world of things/objects which is made by humans.[3] We live in a hybrid world; our humanity is mingled with objects. How can we discover the ways in which the social and the material are entangled together and speak to each other? If we emphasise the agency of matter or the social life of things (Appadurai, 1986), it is the storyline that counts. Objects only become active as they become embedded in a narrative (Harre, 2002).

For Latour, objects that do not move or are 'mute' do not count. In the same way, for Bourdieu, things that are not classified are out of place because the systems of classification are intimately linked to social positions of privilege and hierarchy. For both, agents become part of a story through movement. Movements are, however, in themselves a product of selection. Many systems of scholarly thought find that the mundane situations that occur in the ordinary home settings of family life are often too still, quiet, mute and elementary to be of any consequence. But the stillness of the home is relative to the eye of the observer and to the movement of other entities. There is a lot that happens in front of our eyes that we do not see. Thus, I draw from the practices of scientific laboratories illuminated in the work of Latour (Latour and Woolgar, 1979; Latour 1988a) and the subtle practices of hierarchising tastes and making distinctions developed by Bourdieu (1984), in order to make the familiar and the ordinary strange, or to de-familiarise, in order to bring technologies and families into 'talk'. I use the metaphor of the Russian matryoshka doll to introduce this issue.

In the Russian matryoshka doll model, the small is enclosed by the larger. Latour (2005) uses this metaphor to argue against the practices of classifying agents by size and scales by encasing. Sociology has treated the family as one of the smaller matryoshka dolls enclosed by other more important and bigger dolls. Mainstream sociological theories have only recently come to take the family into account as a major entity in sociological imagination. How does a micro-actor become a macro-actor? Callon and Latour (1981) argue that macro-actors, like institutions, organisations, social classes, parties and states, should not be distinguished from micro-actors, like individuals, groups and families, on the basis of their dimensions. The classification on the basis of relative size between them should be constructed by power relations and networks. However, the usual instruments employed to analyse macro and micro make firm divisions between the macro- and micro-sets of entities like those listed above. Following ANT, the analysis of actors within networks of relations allows for the enlisting to an actor of various entities, for example, bodies, materials, discourses, techniques, feelings, laws and organisations. This means that the more productive way of conceiving the different sizes of social entities – the micro and macro definition – in studies of the family is through connections: the small entity is the one that is unconnected, while the big entity is the one that is attached. The family could be small/micro only if viewed as disconnected from other entities. Some families or households may at different times be micro or macro, depending on the social connections that can be traced from their stories.

If the metaphor of the matryoshka is best dismissed as signifying the enclosing of a small site by a bigger one, both encased in isolation, it can be a useful image for the study of family potentially associated with any other entity or agent of varied sizes, and of sets of matryoshki populating common worlds. The matryoshka denotes a recognisable relationship of a similar object within a similar object, except that inside a matryoshka, a very different one *can* exist, and the possible representations are numerous. The image is significant for the diversity of composition of family units and their connections with each other (see Chapter 2). The social classifications used help to create the world as it is. While each family or household unit can be a set of matryoshki, building on this kind of metaphor we can state that the numbers in each set, their shine, colour and finished style vary. Their closeness to each other also identifies different sorts of classification: they share more common properties.

We are all positioned in households (except for the homeless or those living in institutions) or in families. Households/families thus make up a very large part of the social. But this largesse is significant, in Latourian terms, insofar as associations can be traced between these positions and between these and other entities. Statistical groupings give a shape but not a portrait. In this book I am concerned with recovering 'real people' in social accounts by mixing different materials, generated by various methods, to draw actors and to trace larger connections between humans and things. Yet, I still also retain a very Bourdieusian idea of social hierarchies in a non-flat world where underlying causes shaping the social exist.

Talking with families

A contemporary sociological consensus is that action is distributed among agents and that not all of them are humans (Latour, 2005). When as social scientists we get an account of something, we don't know how many people, or 'things', are present at the same time in this account. There are always unknown and unaccountable agents. In some areas of social sciences, and in particular in feminist studies, the move away from determinist models of social structure, whether technology, capitalism or patriarchy, as the invisible hand directing action in the social generated serious attempts to understand subjectivity, one way among many of counting agents in stories. Although the concern with subjectivity in social sciences has a long history (Simmel and the phenomenologists are eminent exemplars), the status historically attributed to materiality was made highly unfashionable during the 1980s by emerging trends of postmodernism and post-structuralism (Barrett, 1992). Family, personal connections and the investigation of women's lives appeared impregnated with concerns to understand relationality, discursive subject positions and the emotional and affective links of togetherness. Studies of intimacy abounded in this wave in the 1990s. From these perspectives, it might appear strange that disparate themes such as technology, culture and family life should be linked together and given similar status. Yet, my drawing from Bourdieu and Latour intends to make a strong case for the plausibility of their connection within social theory.

Traditionally, the empirical study of the dynamics of family life has been associated in academic practices with demands from researchers for a sensibility to a vast array of problems. If the family were an intimate practice, the experience of researching it would have to be singular. Concerns about forms of eliciting knowledge about different aspects of family life have, of course, depended on the specific academic field (see Gabb, 2008), but the turn to subjectivity has particularly affected sociological approaches. The considerable growth in substantive knowledge and theory about communication within families and interpersonal processes has been confined to a somewhat narrow domain. While concerned with understanding agency, society has been shrunk (Layder, 2004), narrowing down what constitutes social reality. Talking with families has been positioned in mainstream academic areas as a specialist, and relatively less attractive, enterprise than talking with engineers designing roads, social activists protesting against environmental damages, executives producing media programmes, or indeed people using roads or the media. Yet, the attractiveness does not stem from the subject itself, but from how it is treated by those doing social sciences.

Putting together the concerns of studying family, culture and technology challenges both the confining and the special-ness of eliciting knowledge from families.

Technology is a part of everyday life in modern societies and people have complex relationships with machines in their domestic environments. However, as research and development in technology has persistently ignored the everyday, academic studies of the everyday have also often ignored technological developments. Fortunately, there are exceptions. Since the mid-1970s, critical studies of sociology and history of technology have emphasised the complexity and deep embeddedness of technology in the social. Exemplars are Noble (1977) and Braverman (1974) (their precursors include Marcuse (1964) and Foucault (1978)). They refer to technological developments being informed by issues of economics, ideology or religion, all traditions that appear natural and obvious. Some assumptions of this kind have been further developed by ANT and fed into poststructuralist themes informing research about equipment operation on the basis of gender or age, height or skills, and so on. For example, who is a cooker designed for and what tacit knowledge is assumed by the designer when conceiving of the operation of a cooker (Silva, 2000a; see also Chapter 4)? Many ways of life are constructed around a particular technology or ensemble of technologies. The car and the computer are important examples. The washing machine and the fridge (Cowan, 1974 and 1985) are others. However, it is not so easy to realise how profoundly affected by technology our lives are. Likewise, it is difficult for social scientists to devise ways of showing the connections between personal lives and the objects we live with.

Explaining the social is an arduous task. Research informants make their own stories about how the social is made. Following an ANT approach, the social scientist is *not* the one who imposes order, defines the range of acceptable entities or teaches actors to 'reflect' about their practices. In ANT's ethnographic injunction, the researcher follows the actors to find out 'from them what the collective existence has become in their hands, which methods they have elaborated to make it fit together, which accounts could best define the new associations that they have been forced to establish' (Latour, 2005: 12). This implies working with what is not *yet* a sort of social realm. How can the social be traced in studying families, culture and technologies? Family or the home has not figured in Latour's thinking. It appears in Bourdieu's work, but is subordinated to what he sees as stronger fields, its weakness as a field within the social deriving from its strong association with the realm of the feminine. A number of studies on household technologies, mostly informed by feminist perspectives, have developed arguments and insights about families and the social (see Silva, 2002a). These need to be brought into conversation with Latour's and Bourdieu's frameworks to develop better understandings of things and people in family life. I do this throughout the following chapters.

Some of my own biography may throw some light upon some conversation practices: how family and objects 'talk'. When I was 15 years old, I told my mother I never wanted to be a 'hoover driver', like her; I'd rather be a car mechanic, like my father. She was offended – I had wanted to offend her but I felt guilty. I had used labels of attack and desire which, in my eyes, placed my parents in dominant roles, but did not express the work identities of either my mother or my father, for she was a primary teacher, a superb clothes maker (her creative outlet) and he owned the regional coach company. The objects of hoover and car, however, strongly signified the positions I saw them in and driving the latter could take me further. We were a middle-class family of five children (aged from five to 15), living in a prosperous Brazilian town of 50,000 people, with busy lives, and regulated and shared domestic chores. We had a daily maid to help.

In my adolescent eyes, being a wife and mother was simply boring, women's lives were uninteresting, women's conversations tedious, and I wanted out. I still wanted to be a girl, but in a different world, where I would have an exciting life.

One evening, when I was at university in London, I set out to iron a shirt and I felt most miserable: I knew how to do it because my mother had taught me, but I felt deprived since having to do it meant there was no one to take care of me. It had to be me to do my ironing, and I was alone doing it. These feelings sounded like those of an over-pampered, helpless young woman. They were powerful feelings about trivial matters. I felt like censoring my unhappiness. I knew the ironing was a very small thing, but the feeling of 'no one to take care of me' was not minor, and the ironing only revealed the significance of having my shirt ironed for me to give me a feeling of proper care. The trigger of a situation of perceived neglect revealed the mingling of the object of the iron and the activity of ironing with my social position as someone used to consuming paid for ironing services, resonating with Bourdieu's implied naturalness of feelings for 'the material' associated with particular subject positions.

While finishing my PhD, I went to live with my boyfriend in Boston, USA. I came to possess a dishwasher for the first time, but because one needs large amounts of crockery to make use of one, ours remained unused for a long time. Later on we bought a flat in nearby Cambridge, with a kitchen designed by a feminist. I realised the significance of what we had for the way we lived. A kitchen with a sink, cooker, dishwasher, cupboards and rubbish bin within the distance of an arm-stretch, a fridge a couple of steps away, waste disposal built into the sink and an open-plan space connecting with dining room. Ease of work and sociability connected. The flat was in a condominium, and the three flats in the building shared the washing machine and tumble dryer in the basement laundry room. Between home and work we passed two laundry services where shirts could be washed and ironed for one dollar. We used these frequently. All these things were already there in the privileged physical space I came to occupy. This was a social space that also had clear templates for action, delineating roles about the interaction of the machines and our activities. These were largely invisible in a 'seamless web'. However, our ability to engage in sociability with the goods was specific to our historical time, class and gender location, partnership and lifestyle choice. As I stressed earlier, Bourdieu's view of a stratified and differentiated world of social connections, which qualifies the flatness of Latour's vision of the social, does not imply that our social actions as users of the space and machines in our apartment were determined. There was a template for action and we followed it, conscious of some of these templates, modifying some, and not aware of others.

After I had a child I began to perceive that the apartment needed to be cleaned more often. The group of mothers and babies we were part of met in our apartment in rotation, we had more visitors, and babies were soon bouncing on the floor. There was also more laundry to do: baby's clothes, bedding, bath towels and, strangely, shirts that would previously be sent to commercial laundering started to crop up in the laundry basket too. More cooking was done because of the baby's meals, we were going out less and we were having more people in. There was more shopping for food and laundry material, and new things were now needed like disposable nappies and baby wipes. While I carried on working, my work rhythm slowed down considerably. As an academic I was working more from home, and somehow home was making me work for it more and more.

It was at this time that I came across three texts that were to have a profound effect on my thinking. In the order I read them, the first was the book by Ruth Schwartz Cowan (1983), *More Work for Mother*, followed by two journal articles. One article was by Alan Warde (1986), 'Household Work Strategies and Forms of Labour', the other by my then-colleague at Brown University, Hillary Silver (1987), 'Only So Many Hours in a Day'. I entered a world of new conversations (Silva, 2002a). The research subject of the car industry that had been the theme of my PhD work (Silva, 1988 and 1991) was to dwindle and progressively leave room for the study of household technologies and the division of labour in the home. From a car mechanic interest, I turned to focus on hoovers and their drivers, mothering, family life and the everyday world of relationships between women and men, and their children.

My subjectivity and my work were thus fed by the 'intellectual technologies' of the books and articles I read. I happened to encounter them at a juncture where my personal, intellectual and professional life made the texts significant. Inter-objectivity (cf. Latour, 2005) was thus operating, setting the marks of other people (and it is always people who set marks on us, not abstract forces) from other places and times upon my actions. These connections are easy to trace in my own story, but the principle of tracing such connections in my work of studying technology in the family is the same, as we will see in the case studies in this book.

The fragments of my personal journey – emotional, generational and geographical – also highlight a theme that is central to my study. It concerns the role of research in clarifying concrete matters of everyday life, including the significance of personal history, and of the interaction between researchers and researched for the production of knowledge about technologies in family life. When addressing the issue of personal history and the production of knowledge, I am concerned with how a person's story relates to their experiences in the world. Yet, I also believe that I have to follow the stories, and the many agents involved in them, to discover 'real' people. While I trust that accounting for feelings, intentions and so on brings richness to knowledge of an interpretative kind, I am clear that this does not happen automatically. Nor do I wish to disregard the structuring templates of the talk of humans and objects.

How do I go about doing this in my study? I work with life history narratives combining the trajectories, choices and experiences of people living together. Through these, I have aimed to develop an understanding of connections between how people come to live in certain environments, have certain opinions about particular events, and why they choose to do, or are drawn into doing, things in certain ways. The life stories elicited in my study were based on five themes:

1. biographical trajectories, including family of origin and significant transitions in life regarding education, employment, key relationships, spatial mobility and other issues as relevant to particular lives;
2. stories of 'getting together and separating' in developing and constituting 'the family';

3. daily routines in a 'normal' day;
4. accounts of activities like cooking, cleaning, childcare and work, and the roles of technologies in the home; and
5. values and morals (or 'justifications' of action – see Boltanski and Thevenot, 2006) emerging in the context of particular domestic vignettes.

These themes intermeshed in the personal narratives of research participants. For example, a biographical element would came out from a story of how a microwave oven was acquired, or of how a fridge was substituted by another fridge, then a freezer, after a child's birth, illness or a marriage break-up. In another example, a video recording function was learned by one partner but never mastered by the other (creating particular exchanges of favours or conflicts), and the phone bills came to be split after Internet access costs spiralled, and the use of mobile phones was made necessary for a particular reason. With the interviewees, women, men and children, I produced answers using this frame of eliciting stories. I was also an active participant in their home environments for a while. I got people to talk about the issues that interested me. Yet, the diversity of stories produced demonstrates that participants' accounts were not solely constructed by my frames. The frame issues were flexibly deployed, and people elaborated differently on similar issues because their lives are singular, their trajectories distinctive and their desires for the future arise from their individual choices and limits.

However, is it not the point of social science to elaborate models regarding social groups, discover patterns by which certain sorts of experiences are turned into 'social' experiences, and not be restricted to isolated, individual, personal issues? This raises the question of generalisation. What 'structuring templates' can emerge from these individual stories?

To deal with the complex realities of people's lives, my sample design favoured depth rather than breadth and I am clear that there are limits to the generalisation of the cases I studied. At the same time I am sure that the uniqueness of people's stories represents generalisable patterns of the everyday life in domestic settings. This is not to say that these worlds of my investigation are not messy, fluid and contingent, or that I wish to impose an artificially created scientific order upon people's lives (Law, 2004; Law and Urry, 2004). It is simply that in my understanding the world in *not* infinitely messy and that the complexity of heterogeneous realities can be grasped without the imposition of orders. In what ways is it possible to deal with the tension of fragments of data (slices of lives), and the wholeness of a life, in connection with other lives? In order to elaborate on patterns, I have drawn not only on the interviews but also on other 'intellectual technologies' such as data from secondary analyses, archival material, statistical data, historical accounts and readings from media and textual discourses. My aim was also to follow individuals in the material spaces of homes as 'psychosocial subjects' (Hollway and Jefferson, 2000). This meant seeing humans constituted from a combination of unique biographical events, socially shared meanings, interactions and situations. I did this with a particular emphasis: since it is central to my thinking that objects are part of personal relationships, the tracing of associations between humans and things has been my key concern and was my point of entry into the field.

In the next chapter, I look at technologies as they are involved in social change of homes and selves. I do this assisted by segments of particular life stories of individuals and families. I am involved in their stories. In a sense, with them, I carry on with an exercise in autobiography which includes my search to connect technology, culture and family life.

2
Homes and Selves

There has been much concern about the damage that new technologies can bring to the home as mobile phones, pagers, miniaturised surveillance cameras and so on are used from work or other sites to control and reassure that things are in order at home (Urry, 2004; Spigel, 2005; McDowell *et al.*, 2007). How extensively do these impinge on home life? What does the presence of these technologies, and more traditional technologies in the home, mean for sociological thinking about personal identities and relationships? How do homes and selves relate in the context of technological changes? In what ways can these novel contemporary ways of living with technologies be captured in research and how do they challenge theoretical understandings about the relations between material and social life?

In this chapter I reflect on the connections between the characteristics of homes (and houses) and the social practices involved in personal relations that configure identities and subjectivities, looking at the concrete practices of relations between individuals and technologies that are found in homes both in the past and nowadays, to assess the prevailing theoretical claims about these connections. In the next section I discuss theories of technological change regarding the implications they have for understanding relationality in the everyday. This is followed by a description of change in home living in the UK since the early twentieth century to set up the context for an examination of contemporary domestic living. Life in London at the start of the twenty-first century is then considered in the next section, using case studies. The amount of change and diversity of lifestyles encountered show that: (1) transformations over time in the ways individuals relate to one another and the changes in material and social configuration of homes have been considerable; and (2) collective cultural and material resources are patterned and access is differentiated on the basis of social position, some of which are new. These points are discussed in the conclusion, indicating that a refined notion of practice is required to account for individual adjustments within collective socio-cultural and material patterns. In addition, in order to capture differentiated practices within the contemporary diversity of lifestyles and wide access to technologies in the home, the combination of statistics, descriptions and personal stories offers a most productive strategy.

Technology, change and the relationality of the social: a view from home

The world today looks quite distinct from the Britain of the early twentieth century or just after the Second World War, but it has also changed considerably compared to just 20 years ago. A sense of enormous and speedy transformation of everyday life has been expressed by many social commentators. This has been linked to instantaneous worldwide communication and the breakdown of previous local and communal coherences and references. For many contemporary authors like Anthony Giddens (1991) and Ulrick Beck (1992), these movements are supposed to have undermined a sense of a 'place-called-home' (cf. Massey, 1992: 6). For some of those working with families and households, like Roger Silverstone, Eric Hirsch and David Morley (1992), these processes of transformation mean that households are increasingly vulnerable to the threat of outside technological interventions and impositions. It is indeed as if the 'smart home' conception, where humans become integrated into the systems of objects within it, would fast prevail, turning individuals 'thing-like' (cf. Spigel, 2005: 409–11). Assumptions upon which the notions of home life are based and the reality of shifts in home living occur need to be examined in order to assess the theorising about this process of transformation.

I remarked in Chapter 1 that visions of the family as a realm separated from the world at large account for its perceived isolation from the overall process of cultural change. This hinders the understanding of the processes of social transformation, as such an enclosure of the family epitomises conservative views of home life. There is a considerable range of views on this. The ideas of threats, invasions and crises emerging from social, technological and cultural changes and imposing themselves upon the home – and, crucially, the family – are associated with stasis, nostalgia and enclosed security in certain notions of the family and home living. These views separate the home/family from other social locations and regard it as a dependent sphere. What happens in the economy, politics and society, in the world at large, is seen to impact upon the family and to impose itself upon home living. The family thus appears vulnerable and dependent. Changes are seen to occur in *one* direction. Within this assumption, what happens in families (and how home life is lived) is not seen to affect the economy, politics and society significantly. Within these same views, exceptional circumstances are singled out when family life interferes with society. These are such cases as when families produce 'crises', for example, bad mothering creating delinquent children (as criticised by Roseneil and Mann, 1996), or in rising divorce rates because of women not fulfilling their proper family roles (see the critique in Smart, 1996).

Countering such positions, I argue, echoing other feminist analyses, that to account effectively for changes in home life, it is crucial to move away from the assumption that the family occupies a relatively isolated sphere and operates as a system with consensual internal dynamics. There is more 'mess' in family relationships and greater intermeshing with the world at large than these analyses acknowledge. It is important to broaden the questions asked in analyses of home life and break away from an assumed powerlessness of the private. I emphasise the interdependence and circularity of the public and private spheres. While working on a personal level, I am concerned with the collective socio-cultural resources (including technologies) and the patterns of regularities in social interactions. My take on the home is similar to any consideration of place, as understood by Doreen Massey (1992: 12), who argues that 'a "place" is formed out of the particular set of social relations which interact at a particular location'. In this understanding, the home is seen as being constructed out of movement, communication and social relations or, more generally, it is made out of practices that always stretch beyond the boundaries of the home as a location.

Homes are the core empirical space of my investigation and I enter homes interested in the dynamics of relationships involving human and objects, aware that the personal practices I see and that are talked about by research participants are achieved through connections between more than one individual, and that this has been fine-tuned through a long process of accommodation that the person has made to her or his habitat. This involves biographical, material and social issues, and regards both present and past accommodations.

Pierre Bourdieu (1992a, 2005), among others, remarked on how space frames social practices or how habitat and habitus connect. In this connection the built environment may change dispositions, challenging 'normality' or reinforcing it. Practices of power are implicated in the design of homes. Building on a Bourdieusian framework, Kim Dovey (2002) argues that there is a deep complicity between architecture and social order. Framing the habitat of everyday life, architecture structures action by means of construction and positioning of walls, floor, doors and windows, thus constructing the narratives of places in which we live our lives. In this process, the physical space authorises certain forms of identity, the identities of male and female being some such constructions.

In studies about gender and technology it has been shown that in the past, great rigidity of gender identity existed, in contrast with a contemporary present open to greater fluidity of personal and social gender experiences, where bodily experiences are wider and the archetypes of gendered male and female bodies are being challenged, and sometimes transgressed, including with the mobilisation of technology. In different ways these changes affect how people relate with the self and the world over time (Turkle, 1995; Lerman *et al.*, 2002). After looking at changes in demographics and consumption of household technologies, I explore the issues concerning the relationships between selves and homes, considering some of the actions, objects and the propriety of engagement, as assembled in the narratives of individuals in the ethnography, setting out the connections between these assemblages in selected home living experiences. In making sense of these assemblies, I have benefited from the contributions by Laurent Thévenot (2001, 2007; see also Thévenot and Boltanski, 2006) to refining Bourdieu's concepts of practice and the habitus. The contributions by Thévenot and Boltanski are explicitly indebted to Latour's work on the relationship between the weaving of social bonds and the fabrication of objects.

I have noted that practice is about methods of knowing how one proceeds in social life and this is conditioned by past practice, as well as by current circumstances and an imagined future (Bourdieu, 1992). However, as Thévenot (2001, 2007) rightly points out, Bourdieu's theory of practice frequently addresses repetitive and collective types of conduct concerned with social models of behaviour conditioned – if not at times determined – by the collective force of the habitus. Thévenot argues that to address the capacity to shift from one pragmatic orientation to another, according to the demands of any specific situation, including the destabilisation of gender models typical of contemporary societies, practices are best described as: (1) acts that are convenient at the level of the individual; (2) acts that refer to a 'localised dynamic' where there is room for personal creativity; and (3) acts that are bounded by collective dynamics. This makes the personal practice a social practice revealing the self as bounded by the social without, however, stifling the potential for the personal destabilisation of prescriptive behaviour. Any small piece, for example, an individual self or a particular way of arranging a room, is patterned and ordered (see Goffman, 1959) as a profoundly social phenomenon, yet is also individualised.

Transformations in home living and the technological home

Patterns of home living are intricately connected to demographic changes and to choices of partnership and parenting. Technological advances have affected the ways in which homes are made and the demands that everyday life places upon choices of how to live. Table 2.1 shows some of these important movements over time.

Table 2.1   Household types

Household type	1949 %	1979 %	2007 %
Single person	8	23	29
Married/cohabiting – no child	16	27	25
Married/cohabiting – child < 15	37 (there may be other adults)	–	–
Married/cohabiting – child < 16	–	31	21
Married/cohabiting – child > 16 + others	10	7	7
Others	12 (married sons/ daughters, grandchildren)	8 (lone parents) 3 (unrelated adults)	10 (lone parents) 3 (unrelated adults)

No. of persons in household (all ages)	1949 %	1971 %	2007 %
2	22	31	35
3	26	19	16
4	21	18	13
5	13	8	5
6 or more	13	6	2

Source: For 1949: Central Office of Information, Wartime Social Survey (1949) *The British Household*, London: Central Office of Information, in Roberts, 1991, Table A.1 in Appendix). For 1979: *General Household Survey*, London: The Stationery Office. For 2007: *Social Trends*, 2008 edition, ONS, Palgrave Macmillan.

There has been an enormous growth in the proportion of the UK population living alone, an increase of nearly four times between 1949 and 2007 in homes featuring one person. It is interesting to note that while the proportion of elderly living alone has remained relatively stable since the mid-1980s, the proportion of those between 25 and 44 years old living alone has increased rapidly, the number in 2007 being double that of 1986/7. This reflects the trend for later marriage and a higher rate of divorce (*Social Trends*, 2008: 19). There are fewer households with children, dropping from 37 per cent in 1949 to 21 per cent in 2007, and there are generally fewer children in households: the proportion of households with three or more people has declined over time. There is also an increased proportion of families headed by a lone parent, most frequently a lone mother (*Social Trends*, 2008: Table 2.6).

Fertility behaviour, longevity, patterns of employment, rising levels of education and relations between genders account for these trends. These show that the ways people are getting together (or not) to form homes have changed considerably in the last half-century. To build on the notion that home living stretches beyond the boundaries of the home as a location, we can see in addition how the experience regarding technologies in the home has changed in this period.

Marion Roberts (1991) notes that around the middle of the twentieth century, most houses in Britain had electricity, running water, a vacuum cleaner, an electric or gas stove and a built-in bath. However, the census of 1951 also revealed that 38 per cent of households had no bath and eight per cent did not have an outside or inside toilet (Murie, 1983). Adrian Forty (1975) notes that a survey in 1942 revealed that 30 per cent of households earning less than £300 per year (the majority at the time) had no other means of heating up water than in pans on the stove. Alison Ravetz (1995) comments that the impact of the 'wireless' on the home in 1922 was considerable, but television in the 1950s had the biggest effect on home living. How far from these descriptions of homes and home living are the technological contemporary homes?

Table 2.2 shows that electricity, plumbing and gas have achieved quasi-universal ownership in the UK: 95 per cent of households had central heating in 2006 (up from only 37 per cent in 1972) and 73 per cent had at least one car or van in 2002. Television is also a universal consumption good and washing machines and fridges have also reached nearly universal levels of ownership. The spread of consumption of these goods has been very fast. Video recorders, microwave ovens, computers, Internet access and mobile phones have made their presence notable in

Table 2.2 Ownership of consumer durables in households in Great Britain (%)

Year	Wash. mach.	Fridge	TV	Central heat.	Video record.	Comput.	Internet	Microw.	Dishwash.	Tumble dryer	Car/ van	DVD	Mobile
1956[1]	19	7	40	—	—	—	—	—	—	—	—	—	—
1967	61	46	90	—	—	—	—	—	—	—	—	—	—
1972	66	73	93	37	—	—	—	—	—	—	52	—	—
1981	78	93	97	59	—	—	—	—	4	23	59	—	—
1991	87	83[2]	98	82	68	21	—	55	14	48	67	—	—
2000	93	93	99	92	88	45	33	83	26	54	73	—	58
2002	93	—	99	93	89	54	44	87	28	54	73	32	75
2006	96	—	—	95	82	59	—	91	37	58	—	—	80

[1] Data for 1956 and 1967 refer to 'all housewives', which may have had the effect of excluding up to 10% of all households.

[2] Data refers to deep freezers as after 1985, with 95% ownership of refrigerators, data passed on to being collected for freezers only, not for fridges.

Source: For 1956 and 1967: Table 2, p. 23 in Young and Willmott, 1973, on results of a survey carried out for Oldham's Press under the title *WOMAN and the National Market*. For other years, the *General Household Survey*, London: The Stationery Office, 2004. For 2006: *National Statistics*, published 28 January 2008, www.statistics.gov.uk/cci/nugget?id+868, accessed 24 September 2008.

many homes at an accelerating speed. Tumble dryers and dishwashers are also increasingly prevalent, while DVDs emerge in the statistics in 2002, being already present in 32 per cent of households.

However, the picture is far from homogeneous because, despite increases in the overall access to technologies in the home, considerable disparities in individual access remain. Class differences in the ownership of consumer durables were highly significant in the past. They have reduced considerably in contemporary times, with class differences being marked (in both economic and cultural terms) not simply in terms of owning or not owning particular objects, but in the characteristics of the objects themselves in terms of quality, make, style and so on. Whereas in 1956, 42 per cent of all housewives in the 'professional and managerial' class category owned a washing machine, only 13 per cent of the housewives in the 'skilled, semi-skilled and unskilled' category owned one. In 2000, 97 per cent of 'professional' households had a washing machine, as opposed to 93 per cent of the 'unskilled' households (*Family Expenditure Survey*, 2001). More recently, the difference between socioeconomic groups in terms of the ownership of most consumer durables has narrowed even further. In the UK, broadly speaking, everyone has it all. Yet, in the newer communication technologies like home computers and Internet access, and also in particular appliances such as dishwashers, differences in class ownership remain significant. These products are still more likely to be found in the classes of professional and managerial households; the quality of models in these households is also likely to be higher. How does this broad picture interact with personal lives?

Homes and social change

I chose London, a cosmopolitan location, as the setting in which to explore the prominence of claims about how epochal changes, driven by technological change in the wake of the processes of globalisation, affect personal identity and relationships. Supposedly, the effects of globalisation disrupting traditional forms of living in the face of the increasing complexity of social life are felt more strongly in places like London, which are most exposed to these processes. The speed of social change is linked with increased movements of population, technological transformation and availability, and political disputes more prominently found in larger cities. 'Domestic cosmopolitanism' of a high-tech nature (see Spigel, 2005) occurs in London, where the running of homes is predicated on large intakes of state of the art technologies. Yet, I am aware that being cosmopolitan is not dependent on being in a cosmopolitan place. Rather, a cosmopolitan identity depends on articulating the practices typical of a global culture (Castells, 1997b), which are not accessible to all.

Bourdieu relates place and identity in his account of the relationship between social and physical space. His appreciation of the embeddedness of various 'fields' in space is valuable for exploring the significance of identities in social and geographical space, like those that constitute the assets of particular postcodes referred to in the British media (better housing goes hand-in-hand with better schools, better medical facilities and so on). Culturally the ways individuals relate within 'fields' articulate with spatial location, in particular in the field of housing, which continues to be essentially tied to a territorial basis (Bourdieu, 2005). Residential place matters since, as Mike Savage *et al.* (2004) have remarked (in contrast to Giddens (1991) and Beck (1992)), people feel some sense of 'being at home' in an increasingly turbulent world. Thus, the affinity between homes and selves is likely to be more strongly displayed where greater 'turbulence' is to be felt, and emerging lifestyles are also likely to be more intensely present in a location such as London.

I note on the basis of the case studies in London, but also elsewhere, that the diversity of home arrangements for family living has generally moved away from standard styles. Yet, the world is not one of unlimited free choice. Limits to contingent choices are particularly strong when the care of others is involved, and the home remains a very important anchor within contemporary processes of cultural and social changes (on this point see also Chapter 3). In tandem, domesticity has departed from conventional norms. It currently relates to (limited) choices to replenish oneself and one's relationships, this being no longer done exclusively within the moulds of heterosexual housework and traditional housewifery (Silva, 1999b).

As I look at varied pragmatic orientations to everyday life bounded by the specific situation of individual lives, I apply Thévenot's (2001) approach to consider the personal and the collective practices, giving room for creative dynamics of the self within coordinated modes of the social. It is possible to consider the different positions of individuals adapting their home environments to different circumstances: (1) what it is convenient for different people to do; (2) the creative ways in which they use their resources (or fail to use them); and (3) the immersion of individual practices within collective dynamics.

In my sample of 24 families, there are seven London households (see Appendices 1 and 2). All had children of school age: six were heterosexual and were then married; one was a lesbian lone mother; three were Jewish; one was white with mixed-race children (Afro-Caribbean and white), two had mixed nationality (Canadian and English, and German and English), and one was simply white British. Wealth levels were disparate, with annual *net* income[1] ranging from £13,800 (*Lakin*, two adults, two children) to £80,000 (*Churchill*, two adults, three children (H16)).

All homes had considerable access to technologies (see Appendix 2). All but one (*Barker*, the lone-mother household, H22) had all of my selected housework technologies: cooker, microwave oven, fridge/freezer, dishwasher and washing machine. Only the *Lakin* and the *Barker* households had no tumble dryers. These were the two lowest income households. The *Goodman* (H6) and the *Mitchell* (H7) households had tumble dryers but these were rarely used. Rosanne Goodman and Nancy Mitchell had more home-based routines than the other London-based women and they showed greater affinity with traditional housewifery roles. The *Goodmans*, the *Mitchells* and the *Naylors* (H15) had no access to the Internet at the time of the research in their homes (over 1998 and 1999). The *Goodmans* had no computer either.

All families in my sample had greater ownership of household technologies than the national averages. This is because I was concerned with the consumption of technologies in the home and sought to gather those kinds of experiences, although diversity was also important to me. In the *Churchill* household, there was no television, but they had nearly as many computers as people in the home. There are particular stories as to why access to certain artefacts was limited or why people chose to have, or not have, certain technologies. Some of these stories are considered in later chapters. I will now look more closely at four of the London households to illustrate the diversity of contemporary home life against the background of statistical trends and attitudinal changes outlined in the previous section. Three of these households – *Chambers* (H5), *Barker* and *Churchill* – fall into an increasing trend of family types committed to gender equality as far as partnership, employment and domestic divisions of labour are concerned, and one – *Goodman* – relates to more traditional choices of personal style and relationship, which have currently been made by an increasingly small number of people. Let us look at each of them before outlining common patterns and seeing how their individual practices coordinate with the wider social context in which they live.

The Chambers household

The *Chambers* lived in a terraced house in north London, facing a green common, not far from a major shopping area. They owned the property with a mortgage. It was accessible by train and bus but not by tube. The children's school was five minutes' walk away and it took Rose five minutes by car to get to work. Ronald's job was about half an hour away by motorbike, 'a smart and quick way to get to work', he told me, while getting out of his leather bike outfit. They had lived in that house for over 15 years and had refurbished most of it. The house had three bedrooms and a loft that was in the process of being converted into a guestroom. Downstairs, a lounge/media room made a big communal space of leisure and some work-related activities for adults and children (Figure 2.1). The interior of the house was clean and tidy, although piles of things were 'waiting to go to their places', in Rose's words.

Rose Chambers (41) was Canadian, having come to Britain in her early 20s. Work was very important to her for money reasons, but also because it was the way she knew how to live. Family was very important to her, but not traditional domesticity. Ronald (43) shared a lot of the childcare and housework. He was not a 'new man', in the sense of professing full gender role flexibility, he was just a person who made a choice about a partnership with a woman for whom work was important, and he wanted to share the bringing up of the children. He was caring and cared for in many practical senses. Their children, Susie (10) and Steve (6), were well behaved, articulate and sociable. They had various after school activities like drama, swimming, piano, maths tuition and church attendance. Rose normally saw to their attendance to these, but a childminder helped twice a week when she worked longer hours.

The ground floor lounge was a large communal room from which a middle wall had been taken down to increase the space for joint family activities. Music playing, computer use, television and video watching all happened in this space. It was important to Rose and Ronald to be together with the children and the design of the social space of the house responded to this. They were dissatisfied with their kitchen layout which they found to be small, cramped and awkward to use. While hoping to change it in the near future, the conviviality of the lounge space had been prioritised.

The Barker household

The *Barkers* lived in a rented small basement flat in a Victorian terrace in south London, just off a main road, about 10 minutes from a tube station and two minutes from various bus routes. They were surrounded by a variety of shops, takeaways, banks and a post office, in a multicultural

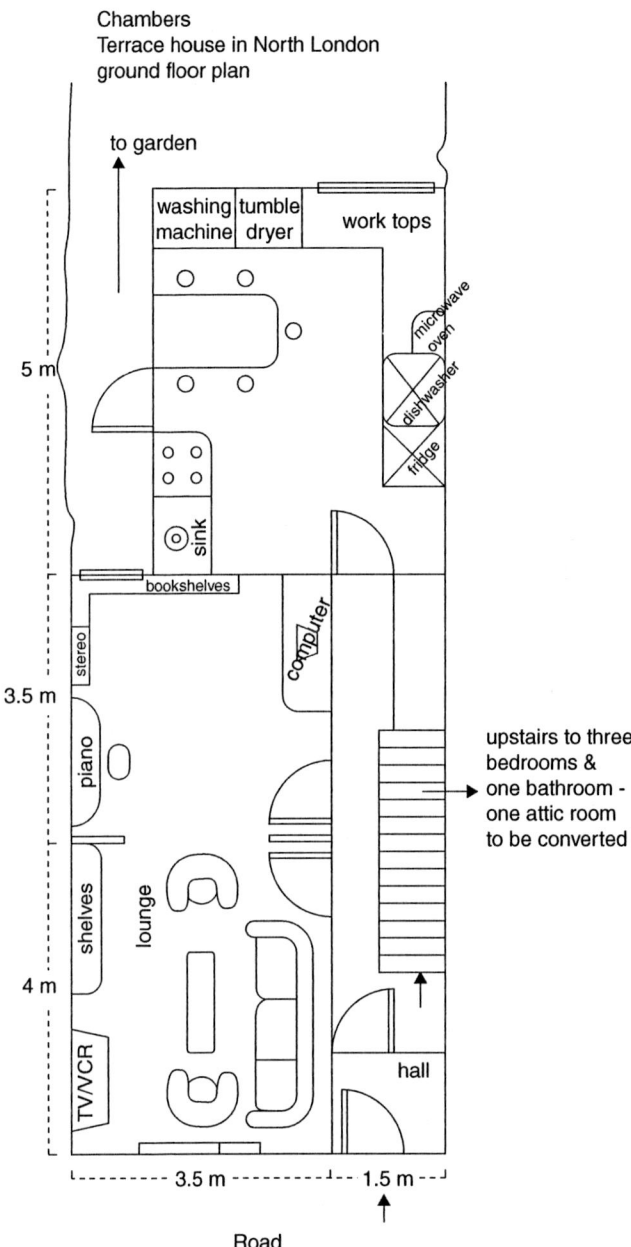

Chambers
Terrace house in North London
ground floor plan

to garden

washing machine | tumble dryer | work tops

microwave oven

dishwasher

fridge

5 m

sink

bookshelves

stereo

computer

3.5 m

piano

upstairs to three bedrooms & one bathroom - one attic room to be converted

shelves

lounge

4 m

TV/VCR

hall

3.5 m | 1.5 m

Road

Figure 2.1   Layout of homes – I

area with a large African-Caribbean population. The flat had a small overgrown front garden. A separate side entrance led to the flat and they had private use of the small back garden, also with an overgrown lawn. The flat was tidy and brightly decorated. It felt comfortable and pleasant though small. All spaces were tiny with rooms having been created out of conversions of previous halls and storage cupboards (Figure 2.2).

Josie Barker (42) was white, but her children were of a mixed African-Caribbean parentage, because of the ethnic background of the co-parent, an African-Caribbean woman, Nadia, with whom Josie had never lived, although they had been partners for 13 years. Both children were conceived by artificial insemination. Josie had had no partner for the last two years, and Nadia remained important in the children's lives and helped with childcare. Michael (11) walked to school by himself, 10 minutes away from home. He had the keys to the flat and got in by himself after school, sometimes with a friend. Josie dropped Cassie (4) at the childminder before going to work and picked her up at about 6.30 pm before getting back home. Michael felt lonely after school and complained about this. Because Josie worked 28 hours a week in flexible time, she could sometimes be home earlier and 'make this effort to be a real mother and cook a real dinner and sit down and eat'. She had a scooter to commute to work and a car 'for everything else'. She had close connections with two other lesbian families with children of a similar age as hers, and this provided her with significant everyday support. Size was a big issue in the house. 'We need a bigger kitchen', said Josie. 'The fridge needs to be bigger', said Michael. 'When the weather is bad clothes need to be dried in my bedroom', said Josie. And Josie was considering moving to a new place in order to get more bedroom space for Cassie. She hoped to be able to afford more soon. Josie and Michael felt they invaded each other's privacy, with Cassie occupying a lot of the social space, particularly since her bedroom fitted no more than her cot.

The Churchill household

The *Churchills* lived in a semi-detached Victorian house, owned with a small mortgage, on a well-off busy street in north London. They were white and British. Their very spacious house felt comfortable. They had good access to public transport but no underground station within walking distance. Other services were varied and abundant in the area. They were the only household in my London study to employ a cleaner. When Diane (43) and Marc (44) started living together, Diane's twin

43

Barker
Ground floor flat in South London

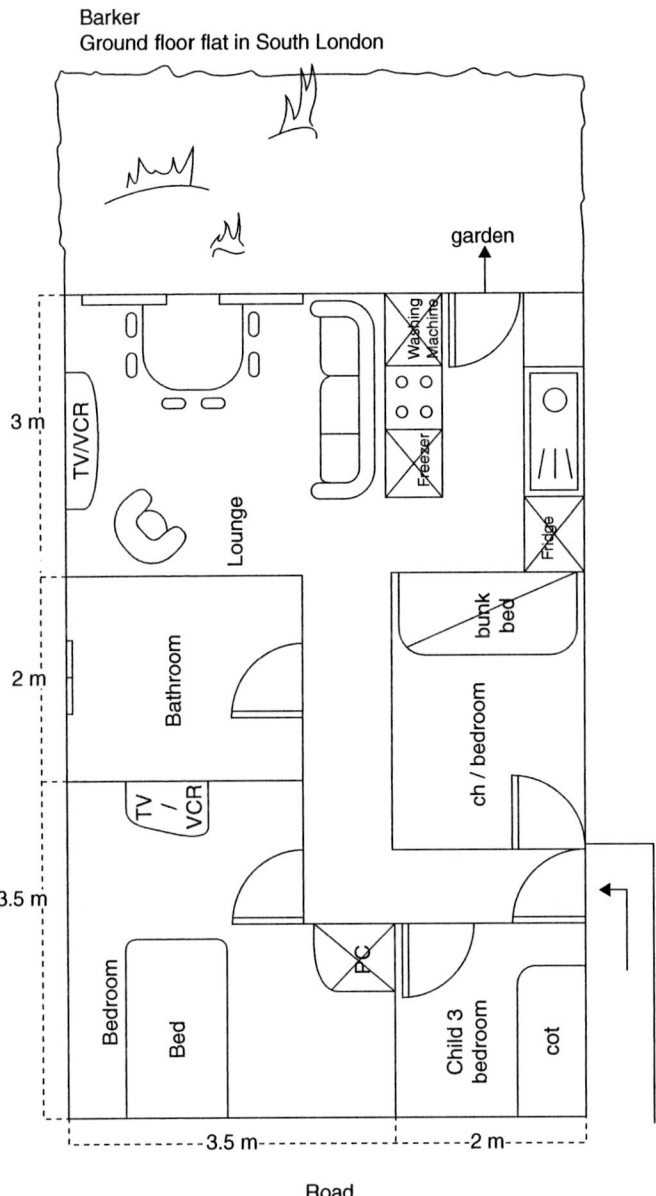

Road

Figure 2.2   Layout of homes – II

brothers lived with them. The four of them bought the house jointly in the mid-1980s. Later on Diane and Marc bought the brothers' share. One of the twins had got married and the other twin, Kirk, still lived with them in a 'bedsit' on the top floor. He had his own bedroom and bathroom. He did his own laundry, worked shifts and was rarely at home. When he was in, he ate with the rest of the family. The children enjoyed his company and they all watched television in his bedroom. There was no television in the house. This was out of choice. They had four computers, had had Internet connection since its early days, and owned a CD video player, which was new on the market at the time. They talked to each other and ate their meals together. They had a car which Diane used for work.

The ways the household worked had a lot to do with how it had evolved. They had a deeply egalitarian way of managing daily life and it was not taken for granted that Diane would do it all. The children were involved in a number of housework tasks: Greg (15) cooked the evening meal twice a week and usually tidied up after meals, while Hannah (11) and Alice (9) unloaded the dishwasher and the washing machine, hung up clothes to dry, dried some in the tumble dryer and checked and distributed clothes on people's beds once they were ready. Having built their early partnership in a household with three adult men and a woman who perceived themselves as equals had made Diane and Marc go through many changes in bringing up children without falling into a traditional husband-wife pattern. The ways they used the space in the house was also singular (Figure 2.3). The front room was a room belonging to Diane and Marc. They went into it every evening after dinner to have coffee, talk and listen to music, or to read. The children only went there if invited. Marc's office occupied the second room downstairs. The kitchen/dining area had been recently refurbished. This was the place for family sociability. It was a very spacious room with plenty of light coming through the glass windows along two of the walls and the glass ceiling at the end of it. A comfortable sofa, upmarket appliances, a wooden dining table and chairs, and fitted cupboards mixed with flowers in vases, bowls of fruit and some books indicated a well-off and relaxed, lived-in atmosphere. Upstairs there were five bedrooms and two bathrooms (with laundry facilities in one of them) on two floors, together with the self-contained flat where Diane's brother lived. They had had au pairs as lodgers before the children got older.

In this house there were designated spaces for particular sorts of relationships. The family related to each other predominantly in the

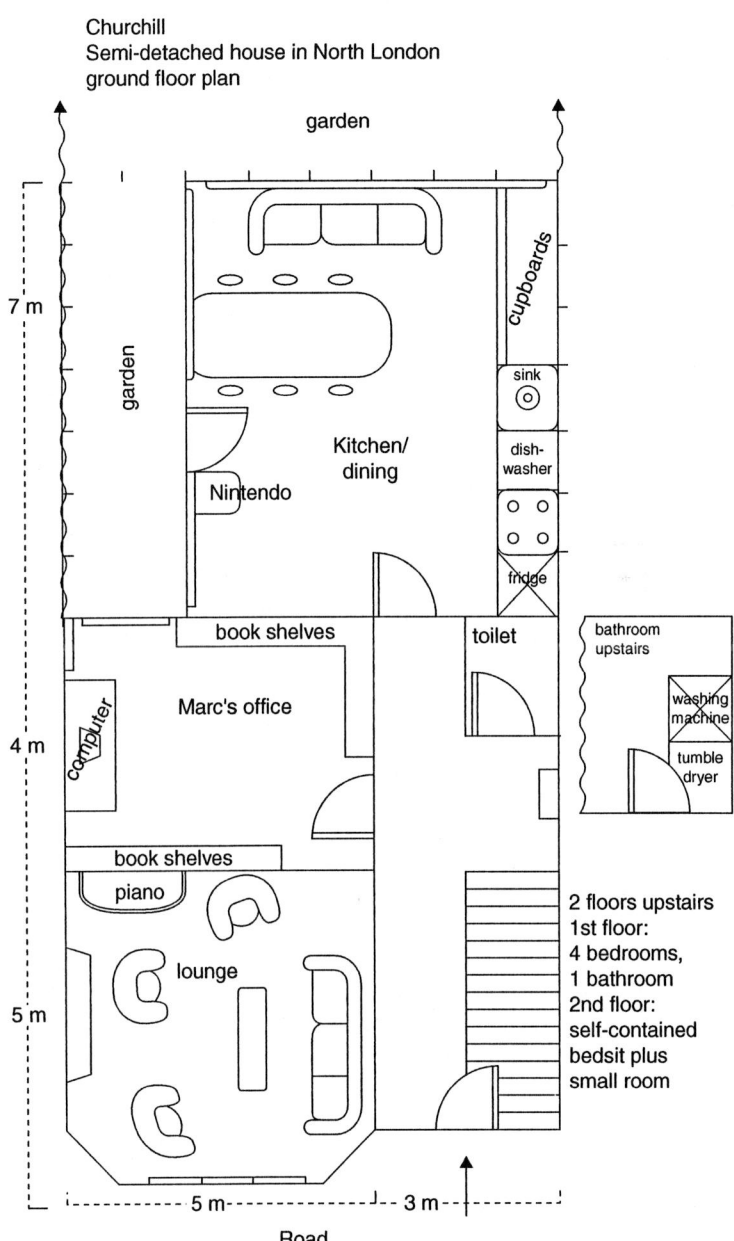

Figure 2.3 Layout of homes – III

kitchen, the couple had their own private room and Marc's work had its dedicated space. There was plenty of space for each and all.

The Goodman household

The *Goodmans* lived in a small detached house in a lower-middle class area of north London, where Jewish and East Asian communities share space. The area had plenty of buses and an underground station 20 minutes' walk away from the house. They owned the property with a mortgage. They followed orthodox Jewish traditions of dress, food and religious calendar, and the children (Eliot, 9 and Tony, 6) went to a Jewish school. The 18-month-old baby was looked after by his mother, with occasional emergency help being provided by grandma. The family owned two cars: the smaller car, which was more than ten years old, belonged to Rosanne (37), while the bigger car with the latest registration belonged to Mike (41). The house had three small bedrooms, and all spaces were small, except for the lounge/dining room area. Everywhere was impeccably clean and tidy. The kitchen area had been enlarged recently, but it remained as a long corridor with just enough space for one person to work in, and a small table, which was pulled out for breakfast. The glass panels on one wall allowed lots of light in and made the small garden visible (Figure 2.4). The large lounge and dining room area was an original design of the 1970s house. It responded to both greater social family living in the evenings, provided a larger space for children's play during the day and enabled the entertaining of guests, while the kitchen and its activities remained tucked away.

I classify this household as traditional, with a husband provider and a wife homemaker. But this identity seemed simpler for Mike than for his wife Rosanne. Rosanne gave different accounts of her routines and work involvement to me and to the research assistant. We came to think that she had some income from work outside the home as a hairdresser and selling crafts. Perhaps Rosanne was avoiding disclosure because of a concern with her tax position, but I think it was because of her ambivalent identity identification that she did this. She seemed to use her odd jobs, which she did not acknowledge having, as an escape from the home. But those jobs were devalued financially and they were an issue of contention in her relationship with her husband. There was ambivalence and she chose to portray herself as a 'lady of leisure' and a responsible Jewish mother with small children, observing religious rituals, which seemed in her eyes to be a more 'proper' way of being. Mike was critical and angry that Rosanne was not at home some evenings because

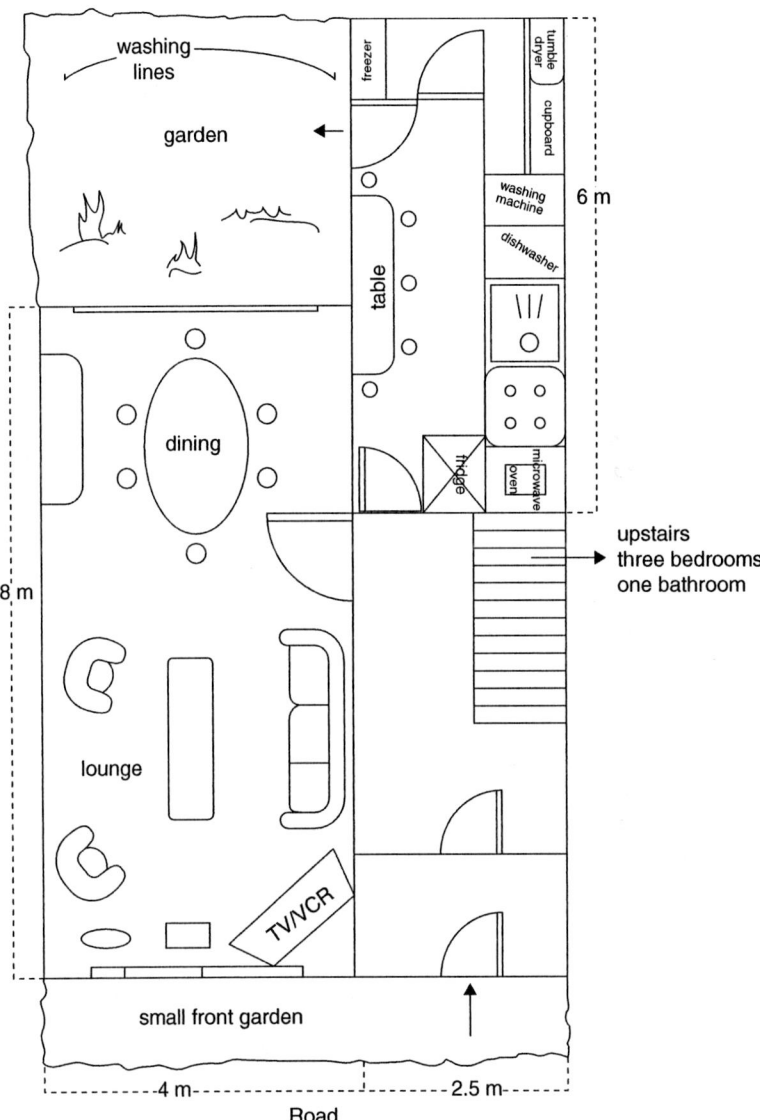

Figure 2.4 Layout of homes – IV

he was tired after a full working day and still, occasionally, had to look after the children.

Selves in homes: changing technologies and identities?

One of my aims in this chapter has been to show how a process of social change configured in the infrastructure of technological consumption over time is linked to demographic trends and the ways in which people go about choosing how to live in the circumstances they make and encounter. This fits with my overall framework for thinking about family life as constituted in open, moving and heterogeneous networks. The ways of making homes are transient and these are intricately connected to personal resources (economic and cultural capitals in particular) and the life course, as well as to various social trends (see Silva and Wright, 2009). The descriptions of the four households show the mingled constitution of the public and private lives of the individuals involved, as well as of material relations embedded in the personal. Arrangements are designed by individuals as part of localised dynamics speaking to the convenience of choices and the creative organisation of lives (Birdwell-Pheasant and Lawrence-Zúñiga, 1999; Thévenot, 2001). Yet, in the terms devised by Bourdieu, collective dynamics clearly have a role in binding relations and in personal choice. These include housing structure and affordability, job flexibility, occupational roles and income as much as the possibilities and desires to live gender roles and sexuality in specific ways.

The cases also assist with my other, but closely linked, aim in this book, which involves thinking about the connections between the personal and the social, and of reworking the relationships between individuals and the material environments they occupy. For example, this relates to the sense that relations external to the person, like their income and their job, as well as the houses they live in, come to be seen as intrinsic to the relations they have with other people.

Although by no means deterministically, there is a connection in these four cases between the low income and intermediate class position of Josie Barker and her having the smallest housing, despite her university education, which was at an equal level to that of Rose Chambers. Her alternative lifestyle, as a lesbian lone mother, has a bearing on this position. The partnered relationships of the other three households account for higher income levels and larger housing affordability and, to some extent, the heterosexual couple partnering matches the lifestyles displayed.

The highest income household, the Churchills, had both partners in professional executive occupations, who were also holders of high levels of cultural capital, as displayed by levels of education and participation at cultural events (see Chapter 5). They had the most expensive and largest house. They dealt with gender roles in mobile and flexible ways, not only between the adult partners – Marc was at home nearly full-time while Diane worked outside of the home full-time – but also regarding the children, with Greg, like his father, being in charge of cooking, while all of them shared with the laundry. The Chambers household was equally in the professional executive class, with high cultural capital, enjoying flexible gender roles, despite not being as progressive as the Churchills. They benefited from employment with flexible hours and both parents cared for the children and for one another. At the Goodmans, the inter-mediate occupational class of a low-level manager with a full-time mother and housewife goes together with smaller housing and an assured choice of traditional gendered domesticity. However, greater certainty was conveyed by Mike about his preferred gender arrange-ment than by Rosanne who, in full charge of childcare and housework, showed some disquiet in appreciating her role. She appeared to fit in a predetermined social contract as a result of being in it, not because of seeking or desiring it. Of course, there are women and men who want and convincingly pursue traditional gender roles and these are not con-fined to lower occupational classes, even though this was the case with the Goodmans.[2] These four households illustrate varied ways of living, but they do not encompass the full range of possibilities.

The centrality of gender in these arrangements is evident. The cases show that masculinity and femininity can be performed in the home and mobilised and contested in different ways, offering different rewards regarding social position. Some women want to work as part of their identity. Some men want to look after their children and, sometimes, after their homes. Childcare practices of lone mothers have a higher reliance on commercial arrangements and friendships. Some men and women practise traditional domestic gendered ways of organising everyday life which, on occasion, may generate ambivalent identity identifications. Within this empirical diversity there are some broader theoretical issues that underpin the ways in which we can see the operations of gender as manifestations of the self in the context of home life.

There is an assumption, prevailing in the social contract of recent times, that gender in the family is unproblematically a personal and pri-vate matter. This explains fieldwork practices of interviews of 'couples' in family, as undifferentiated individuals, oblivious to the differences between partners, as is the case in the study by Silverstone, Hirsch and Morley (1992). This assumption negates both the interrelatedness of doing gender and the relationship between the public and the private in the person and in the family. Gender is performed in context and with other persons. What goes on in the world at large, for instance, the ability to work flexible hours, earn a large income, take decisions and deploy various gendered assigned characteristics in diverse ways (men doing femininity, women doing masculinity), relates to arrangements in the home as much as to the bearing that relations in the home have on the wider world. Gender, as a property of the person, is in the current socio-cultural British context a potentially flexible matter.[3]

The four households illustrate ways of living family life which, with variations, portray situations of possibly about 20 per cent of the British population – those households with children younger than 16 – at any one time. The proportion is, however, much larger over the life course of the population as a whole. Most people in the UK nowadays, despite increasing childlessness, pass in and out of this category of 'families with school aged children'. However, there are varied arrangements within this group. Women working in outside employment while being mothers of school-age children is a growing trend, even if this employment is mostly with reduced hours to allow for them to perform their childcare role. There are men who hardly do any childcare or housework, like Mike Goodman, but Ronald Chambers and Marc Churchill are heavily involved in the daily lives of their children and home. Both statistically and in my ethnographic study, the case of the Goodmans appears to resemble a pattern that prevailed up to the 1960s but which has been numerically and attitudinally reduced. This type predominated under the social contract based on a strong gendered division of labour. As Lisa Adkins (2005) notes, this sort of contract assumes that skills and expertise are attached to, and accumulated by, the person, and these properties became organised internally as self-identity, self-esteem and bodily capacities. Rosanne, and 'proper' housewives in general, even if they are also workers, are assumed to do femininity of a traditional kind. I explore this further in relation to technological developments in cooking and cleaning technologies and practices in Chapters 4 and 5.

Appropriately, Adkins (2005) notes that in the 'new economy', gender-related properties are no longer stuck to the person. Gendered properties are organised externally. For instance, it is not women who have femininity, but femininity becomes a skill required of certain job positions, regardless of the gender of the occupant. I argue that these patterns are extensive in domestic life. As exemplified by the cases of

Diane, Josie and Rose, mothering and housewifery are not properties necessarily attached to women, mothers or wives. The skill to mother, or the expertise to manage the home, has to some (albeit small) extent become increasingly detached from gender, and we see men engaging more in these traditionally feminine domains. Creative adaptation to localised dynamics shapes the individual and collective practices in the terms outlined by Thévenot (2001).

However, social theory, like that of Bourdieu, explored gender trans-formation as being trapped within the traditional social contract model of personhood from which masculine domination emerges. This has major consequences for his view of the 'normalcy' of families, a pattern largely dissonant with family life in current times (Silva, 2005). The patterns we see in this chapter, where individual creativity and collec-tive dynamics are bound together, show individuals creatively adapting technologies available to their ways of relating. Interestingly, no sense of damage or threat from technological changes was evoked. Domestic lives are capable of allowing individual capacity to flexibly address collective orientations. An important theoretical assessment indicated by my empirical material is a call for refining the Bourdieusian notion of practice. I endorse Thévenot's claim that individual practice can be a lens for seeing what becomes invisible within collective practice. Yet, because subjectivities contain the social, I can further claim that, methodologically, individual case studies offer refined means for capturing practice and theorising about the social.

In the next chapter I explore resources of time in practices of domestic routine and in the development of household technologies, and I further develop the argument about the connections between the material, the relational and the cultural, indicating the social coordination and patterning of 'messy' everyday life.

3
Time

Social processes happen over time but they also happen all the time. In this chapter I argue that people's routine practices are of great importance for broader movements in society and that relative uses of time reveal the interrelation of both subjective and materially informed bases of social life. One of my concerns is to focus attention in social and cultural theorising on making ordinary uses of time prominent. I am mindful that reflections about time in everyday life figure in recent sociological work (Bourdieu, 1992a; Giddens, 1991; Beck, 1992). Yet, despite the recognition that routines matter, concerns in this mainstream theorising remain somewhat abstract, with the most routine aspects of daily existence, like body maintenance, emotional nurturing and so on, not being seen as a relevant focus of grand theories. These have, however, been prominent in academic feminist perspectives since the middle of the 1970s. For instance, Dorothy Smith (1987: 81) remarked that it has been taken for granted that these ordinary matters are provided for in a way that does not interfere with social life in general, and thus will not affect conceptual models. The invisibility of these matters is linked to practices which position the subordinate groups as the carers for bodily and materially located needs (see also Tronto, 1993; Sevenhuijsen, 1998). Household chores, eating, dressing, reading, getting to work and so on have generally been regarded as unproblematic matters in social theory. More recently, Lisa Adkins (2009) has claimed that the centrality of women's roles in domestic labour is at the core of the contemporary capitalist relationship with productive labour. If women's reproductive labour in conditions of Fordism was not measured because it involved cognitive, interpretive and relational activities, these same sorts of activities have become integral to current productive labour in the wider economy. The implications for theorising gender are profound and most significant in the

UK where, regardless of domestic commitments, the majority of women are in paid work.

Time, as a complex concept related to sociological knowledge, raises questions about what counts in temporal analysis, how the counting is done, what is the value of time and whose time counts and is valued. While I pursue these issues in this chapter, I further develop them in subsequent chapters, discussing particular activities like cooking, cleaning and domestic dilemmas. Issues of social change both in industrial production and in social reproduction demonstrate that analyses of time use entail difficult methodological and theoretical tasks (Adam, 1995). When sociology and other contemporary social sciences address practices of our everyday world, traditional methodology is turned on its head. Rather than being guided solely by abstract concepts, concrete practices need to be traced and followed for interpretation, in the manner of Latour (2005). Time is in itself a category that allows for diverse interpretative stances. It entails various different representations of the world and can be classified in various ways. For instance, ordinary home routines may appear extraordinary to those who lack certain kinds of everyday experience.

In this chapter I engage with these debates through an exploration of two key issues. In the next section I consider the ways in which theories about the development of household technologies have engaged with changes in the rhythms of family life. This is where views about the conservative 'impact' of technologies on social relations, in particular on gender divisions in the home, are evoked and challenged. I take issue with these views on methodological and empirical grounds. In the following section I explore data – statistical and narrative accounts – concerning everyday domestic routines, in the context of changing working and relational lives in contemporary Britain. In concluding, I reflect on the importance of emerging patterns where for increasing numbers in the population (chiefly comprising women but gradually spreading to larger groups of men), the differentiation between working time and non-working time, time for care and for leisure, technologies for work and for pleasure, have become empirically hard to discern, challenging theoretical understandings about boundaries in significant relations in people's lives.

Household technologies

I have been pursuing the idea that homes and family life are active contexts operating in interaction with other social contexts, claiming that personal choices about how to live may make a strong and active mark on wider economic and political contexts. In this book I show that new ways of organising the domestic everyday have evolved in conjunction with the household technologies that are embedded in everyday lives, which are however elusive. How is the deployment of household technologies linked to social and personal time? How does it link to culture, the economy and politics?[1]

The 'classic studies' of household technologies informing debates in the 1960s and 1970s (Bose, 1979; Cowan, 1983; Thrall, 1970; Vanek, 1973) noted a continuing paradox that the amount of time spent on housework seemed impervious to technological change. In addition, they asserted that the gender division of labour remained unchanged in the face of technological influences. The explanation focused on emotions: a good woman/wife/mother labours with love, and standards of housekeeping were signifiers of this love. Within a patriarchal ideology, standards have increased to offset the time saved by technology. There is always 'more work for mother', according to the title of the book by Cowan (1983), which so motivated my interest in household technologies and my need to challenge this assumption. The assumption embraces the domestic stasis (while change happens elsewhere, the domestic remains stable) also noted by Lynn Spigel (2005) in relation to current technological developments of the 'smart house', by which domestic hierarchies are inexorably reasserted in the processes of change envisioned by the technological innovators.

Let us place the discussion in a wide historical context of which homes are a part, like that of the process of classic capitalist accumulation, where the centrality of labour time translates into work efficiency, derived from a basic equation between effort and time. Effort is measured in terms of human motion within a particular timespan. Nowhere was this more developed than within Taylorism, which, in the early years of the twentieth century, sought to establish scientific principles of management based on measuring workers' motions in varied work situations to establish rules and methods of work organisation involving minute operations. Fordist models of production adopted Taylorism to respond to worker restrictions on employers' use of their time, via a compression of 'wasted' time. In the 1920s, Taylorist principles were also applied to domestic management by Christine Frederick (1919), a home-efficiency expert. This pattern of time intensification had a number of consequences for the organisation of private life, chiefly the roles of homes and women, but also of housing designs and the 'imaginary relationships between individuals and objects' (Aglietta, 1979: 161).

The Fordist model assigned a special place in society to the female population. Of course, housewifery had already existed for about four centuries (Hall, 1980), but Fordism assigned women the provision of housework in the nuclear family, reconstituting wage-labour. As Fordism was based on individual ownership of commodities, the role of women as consumers was enhanced. Small family units, standardised housing and the motor car were related to this pattern. Standardised, chiefly suburban housing was constructed according to certain basic standards to put an end to unhygienic and unsafe interiors and to enable the installation of household appliances that saved domestic labour. Practices at home were then changed with the use of time-saving equipment. This process, however, had to combine the use-value of household equipment with capitalist mass production. An overall set of social relations were combined to closely link the labour process to the social consumption norm. Emerging around the 1920s, this pattern was particularly marked in the 1950s and 1960s. Fordism implied a distinct formal separation of the location and roles of men and women, private consumption, and the preservation of the home as a 'haven' for physical and emotional replenishment for the predominantly male workforce. Male labour had to provide earnings high enough to enable the purchase of consumer durables and equipment for the home, and to allow housewives to stay at home in order to perform the caring activities required by husbands and children (as consumers and producers of the future). The Fordist model was based on an unequal interdependence of the conjugal couple and on women's lack of autonomy (Lefaucher, 1995). The 1970s feminist debate on domestic labour focused on this particular social dynamic and revealed the hidden disadvantages for women (Gardiner, 1997).

In a post-Fordist context, like that in my investigation, families are positioned to provide for the satisfaction of material and emotional needs (even though not all families do this). Families are or are not transformed according to whether they continue to serve these needs. Of course, this is not an unconstrained or inconsequential choice. Personal and family scripts are written in the context of different social and economic locations of families, as well as individuals, within wider social structures. They also vary according to the available emotional, cultural and economic capital possessed by each individual, in terms of age, gender, professional qualifications, position in the labour market, sexual preferences and parental obligations, among other factors.

The 'classic' studies of household technologies in the 1960s and 1970s were trapped in the social theorising of Fordism. They reflect a conception of segregated and compartmentalised time of a sacredly private place for homes and families. They conceived of a sharp boundary between the private and public spheres. The more recent studies of domestic technologies show some shifts in work time and gender roles in homes, a trend that increased towards the end of the twentieth century. In the post-Fordist model, such changes have important implications for the ways in which everyday time and care are dealt with in families.

Recent research on household technologies has broadened from a focus on housework in the home to include the roles of industrial design and manufacturing in shaping the consumption and gender of consumers of household technologies. It has also been more sensitive to cultural diversity (Cockburn and Ormrod, 1993; Cockburn and Fürst-Dilic, 1994; Chabaud-Rychter, 1994 and 1995; Silva, 1999b, 2000a and 2004). In these newer studies, it has been argued that the causes of the non-reduction of household working time were to be found in technological manufacturing strategies and wider patriarchal power relations. The main arguments are that few domestic innovations were designed with the intention of reducing housework time and fewer innovations were designed to change the gender of the houseworker. Empirical research demonstrated that manufacturers of household technologies had aimed at enhancing the activity to which the appliance was connected, instead of making the activity easier, simpler, quicker, non-skilful and non-gender related. The aim of industrial designers and manufacturers was to *improve* female housework. This was a consequence of the power positions of men in manufacturing industries, which have given them scope to affirm the traditional gender division of labour, defining and shaping the work of the contemporary woman. These arguments are reminiscent of those relating to the impact of the prevalent patriarchal ideology upon household technologies found in the 'classic' studies. But there are notable differences, particularly in the exploration of the complexity of changes in gender relations.

Women and housework have traditionally not been highly regarded by designers and manufacturers of household technologies. For instance, in design engineering processes, it was found that male designers introduced women into their labour processes in many disparaging ways.

They either put themselves in the place of the woman (Chabaud-Rychter, 1994), engaged women as testers (Cockburn and Ormrod, 1992), hired women as assemblers while making them stand for the housewife or tried out products with their own wives (Gomez, 1994). Reliability tests for various household technologies were devised, imagining a woman user who was careless, clumsy, absent-minded or dangerous in using appliances (Chabaud-Rychter, 1995). Around the turn of the twentieth century, these 'imaginary relationships' (Aglietta, 1979) still appeared as incongruous Fordist reminiscences.

The places allocated to women in the subjective minds of those involved in design, manufacturing and production, as found in these recent studies, reveal an ingrained perception of women as subordinate in a world segregated by gender, which still evokes the Fordist model. Obviously, modes of social organisation change only gradually and unevenly, but in the early twenty-first century, the gender order has changed significantly in comparison to that of the middle of the twentieth century. By the end of the last century in Britain, for instance, the number of families living exclusively on a man's wage had dropped significantly and fewer people accepted that a wife's job was solely to look after the home and the family, even though most people still expect that women should perform more of the household chores (Scott, 2008). The newer studies also found evidence of new technologies being accompanied by more equal gender relations in the home, where some men had greater (though still relatively little) participation in housework.[2] One example was found in England, where men appeared marginally more involved in 'cooking' because of the microwave oven (Cockburn and Ormrod, 1993). Another example refers to Finland, where men appeared to be doing 10–15 per cent more vacuum cleaning since a new appliance was introduced (Smeds *et al.*, 1994). I pursue these arguments about gendered participation in cooking and cleaning in the next two chapters.

How do changes in the lives of individuals and their choices of how to live interact with demands for technologies for the home? I showed in Chapter 2 (Table 2.1) that in Britain in 2007, only 21 per cent of households had a woman, a man and children, while 10 per cent had a lone parent and 29 per cent were one-person households (*Social Trends*, 2008). Similarly, the proportion of economically active women has been rising (see the next section of this chapter). What are the implications of these various different home contexts? What is to be done with homes when women are increasingly in outside employment? Who will care for the home and family? How is this caring to be performed? These are the questions of a post-Fordist time, and the answers involve a changing picture of the roles of women and men in families, of their gender-related properties both in the labour market and in the (technological) home, combined with how time is disposed of as a resource.

Childcare is obviously a major problem, and the growing rate of employment among women has been linked to later childbearing and part-time jobs. In the late 2000s, the majority of working women in both full-time and in part-time employment had children, and the employment of those with younger children had increased. Since the mid-1970s in Britain, the economic activity rate of married and cohabiting women has risen steadily. Also, there was a continuous increase during the 1980s and 1990s in the economic activity rate of women with dependent children, especially among those with children under five. Similar trends are found in the USA, although with a smaller number of women in part-time jobs. There are several ways of looking at the implications of these trends.

One approach emphasises the 'flight from home'. Arlie Hochschild (1997) has argued that in the USA in the 1990s, both women and men were fleeing into work from the tensions at home. Relationships at home grew strained due to women's lack of time for emotional work, and such difficulties had increased in a context of family fragmentation: second families, reconstituted families and former families, as well as in intact families. The resistance of husbands towards 'helping' was another major source of conflict. This family life scenario contrasts with that at work, where the individual has been increasingly valued and cared for within employee-centred management programmes. Yet, tensions increase in the home because caring work alone does not provide enough emotional nourishment. Together with the issue of lack of time, it seems that there is a strong need for an 'emotional culture at home'.

Another way of looking at the implications of the changes in the ways in which people are currently living their domestic lives emphasises the role of the market and economic imperatives (Schor, 1993; Silver, 1987). Juliet Schor (1993: 98) has argued that 'housewifery is dying out'. This means that fewer households nowadays can afford the labour of an adult solely to undertake housework, cooking and mothering. Dual-earner households predominate in post-Fordism, within a different 'wage relations norm'. In past conditions when domestic labour was cheap, there was not much incentive to save it, but when housewives' time is at a premium (because of economic, social or cultural reasons), households have to begin to behave differently. Economics does not explain the whole story because, despite the noted reduction in time spent on housework that accompanied increased female labour market participation, Schor argues that women are still locked into a model of household technology use and a culture of domestic work that are inefficient, time-consuming and onerous. The cause lies in the low cost of the housewife's labour. However, this becomes more expensive as women spend more time in the labour market. In Schor's view, families 'bought' so much domestic labour because it was very cheap, and standards and the amount of services escalated. Conversely, as women's income began to count, the time women spent on domestic labour began to shrink.

Hillary Silver's (1987) work complements and refines Schor's arguments. She says that the value of women's time only counts insofar as it enables them to buy replacement domestic services. In her view, time spent on domestic labour can only be reduced by access to the labour of others. Women can either buy other people's time or force other people, through the state, to share the burden of domestic labour. Both these strategies involve demands on the service industries. The tendency is for a transfer of housework from the private home to the public space of the market. This is why the growing industrialisation of housework in the USA in the 1980s was associated with the growth of employment in specific industries such as laundries, eating and drinking, paid childcare, repair services and hotels.[3] Technological improvements affected housework only via innovation in the service industries, with the implication that housework would be displaced from the home. Silver does not consider changes in the gender division of labour in households: she presumes that, since men do not do housework, it is virtually impossible to change the domestic gender division of labour.

However, others have found that the proportion of total household domestic work undertaken by men has risen. This emerges not just out of choice, but from a process of 'lagged adaptation' (Gershuny *et al.*, 1994), in which the adjustment of gendered work roles in the home takes place through a process of household negotiation over a long period of time. Contrary to the 'classic' studies of household technologies, and in line with other findings from more recent studies, the time women have spent on domestic labour in their homes indeed appears to have shrunk (Gershuny, 2000; see note 2 in this chapter), though the overall time people spend on housework has recently been reduced. For example, in 2000, 86 per cent of men and 96 per cent of women spent some time on housework, compared with 77 per cent of men and 92 per cent of women in 2005 (Gershuny, Lader and Short, 2006). The logic of male-female exchange has been marked by increasing commodification.

In an earlier study, Jonathan Gershuny (1983) noted different patterns for working-class and middle-class households. He found that for working-class women, time spent on housework increased slightly from 1930 to 1950. His hypothesis is that the early stages of the development of appliances may have led households to engage in *more* housework. However, time spent on housework decreased quite sharply from 1950 to 1980, possibly because the efficiency of domestic capital increased. Middle-class housewives in 1930 did about half of the amount of housework done by working-class housewives. However, the time spent on housework by middle-class housewives increased sharply up to 1960, partly because the loss of servants increased *their* time spent on housework. The difference between working-class and middle-class housewives was then minimal. But, from 1960 to 1980, time spent on housework decreased markedly. This reduction, both for working-class and middle-class housewives, has continued, given that women who are employed full-time tend to spend less time doing housework than the diminishing number of full-time housewives. In the more recent studies of household technologies, I noted earlier that it has also been found that men, albeit still in small numbers, become more involved in housework, childcare and cleaning (see also Gershuny *et al.*, 1994; Sullivan, 2000), particularly when certain kinds of appliances are available. Interestingly, however, this has increasingly been compounded by the creation of new social divisions between women in practices of childcare and domestic labour provision by other people not related to the family (Lutz, 2007; McDowell, 2008). The everyday routines of the families in my ethnographic study allow for a detailed exploration of current patterns.

Everyday routines

The greatest post-War change in the use of time in the Western industrialised world has been the increase in female labour market participation. In 1971, 91 per cent of men of working age, compared with 57 per cent of women, were economically active in Great Britain. The activity rates of women have generally increased since then to reach 72 per cent by 1997, while the activity rates of men slowly declined, reaching 85 per cent in 1997 (*Social Focus on Women and Men*, 1998). By 2007, the gap between the numbers of jobs held by women and men had reduced to a very small difference. Although women were more likely to work shorter hours, the total time worked by men had fallen in absolute terms (*Social Trends*, 2008: 52–3).

What are the consequences of such an increase in female economic activity rates for the uses and structures of time in society? And how do these changes take shape in people's daily lives? Some of the theories I outlined in the previous section advance enlightening explanations. The woman in the household who acquired a new washing machine is now more likely to have a job than was the woman in the mid-twentieth century, who could have called on the laundry service. Even though she now has the technology, she is likely to be the one doing the laundry work. Paid and unpaid jobs cumulate on women. Yet, many of these jobs and the dynamic of activities between them, like the fragmentation in the use of time of the woman doing the ironing while keeping an eye on the children playing nearby, daydreaming, listening to the radio and planning the evening meal, are issues of time that require detailed investigation.

In Figures 3.1a and 3.1b the patterns of routines in the everyday lives of 42 individuals (24 women and 18 men) in 23 households are represented.[4] These are visual representations of the cycle of the waking day of particular individuals. Their routines are located in culture, history and personal biographies. To generate this data, I was concerned with getting people to talk about what they did, the things (technologies) they did things (activities) with, and about what happened as they did these things. I wanted to obtain in-depth multiple narratives of their everyday lives.[5]

I generally asked people to tell me how a 'normal' weekday would go by from the time they woke up until they went to bed. The manners in which people kept track of their everyday narratives were quite diverse. Some lost themselves in sidetracked accounts of events triggered by a recollection of some occurrence in the everyday. Normally men were either very brief or concentrated on events at the workplace. Because I was interested in what happened in domestic life, I encouraged people (most often men) to tell me what happened in the home. I also encouraged accounts of people's perceptions of the routines of their partners and children. Again, for men this had to be prompted more often than for women. Women's routines appeared more generally connected with those of their children in particular, but also with those of their partners.

Because some people's daily lives have different set patterns for different days of the week, we would normally talk about these differences.

Time, space, activity and people appeared as recurrent elements in the narratives of routines. Time routines were constructed around activities performed in particular spaces, involving specific individuals. The common ingredients of a 'normal' routine included getting up, washing, having breakfast, giving children breakfast, taking children to school, childcare, setting off to work, commuting, doing housework (cleaning, cooking, washing, ironing), working, having personal time, engaging in relational time and watching television or spending time on the computer or the Internet. The activities most often interlinked with other concomitant activity (or activities) were childcare, housework, watching television, having personal time or engaging in relational time. Work was the most preserved single activity done in isolation.

The data in Figures 3.1a and 3.1b shows that eight people, five women and three men, spent practically the entire day at home. However, their experiences of home life differed significantly in terms of employment position and work interests, the age of the children in their care, and to some extent in relation to their class positions. Gender is a strong differentiator, but it is dependent on negotiations and bargains for uses of time in and out of the home which are divided up between partners according to their resources and desires.

Tracy Green, in household 1 (H1), was an architect working from home. She got up at around 7 am, with her children. She worked while the children (who were five and eight years old) were at school, and again most evenings. Housework and childcare were mostly done concomitantly. Katie Hughes (H2) normally got up at 6 am, around the time that her husband, working as a builder, also got up. Her day had personal time, childcare and housework, with the television on all morning. After lunch, while her younger daughter slept, she had personal time again for a couple of hours, followed by more childcare and housework, broken up by some after-school activities. Her evenings varied, as she worked some days at the nearby supermarket. If she was at home, she spent some time with her partner and also had some personal time watching television. Wendy Bird (H9) got up early, with her husband, an executive professional. She did much more housework than Katie, who combined housework and childcare for most of the day. Her children, being older, went to school. In the evenings she worked, from home, for an hour and a half, giving tuition and earning some money. Evenings also combined childcare and housework with television watching until 9 pm, when the children went to bed, after which she carried on watching television on her own or with her husband. In household 10 (H10), Rena Rock was

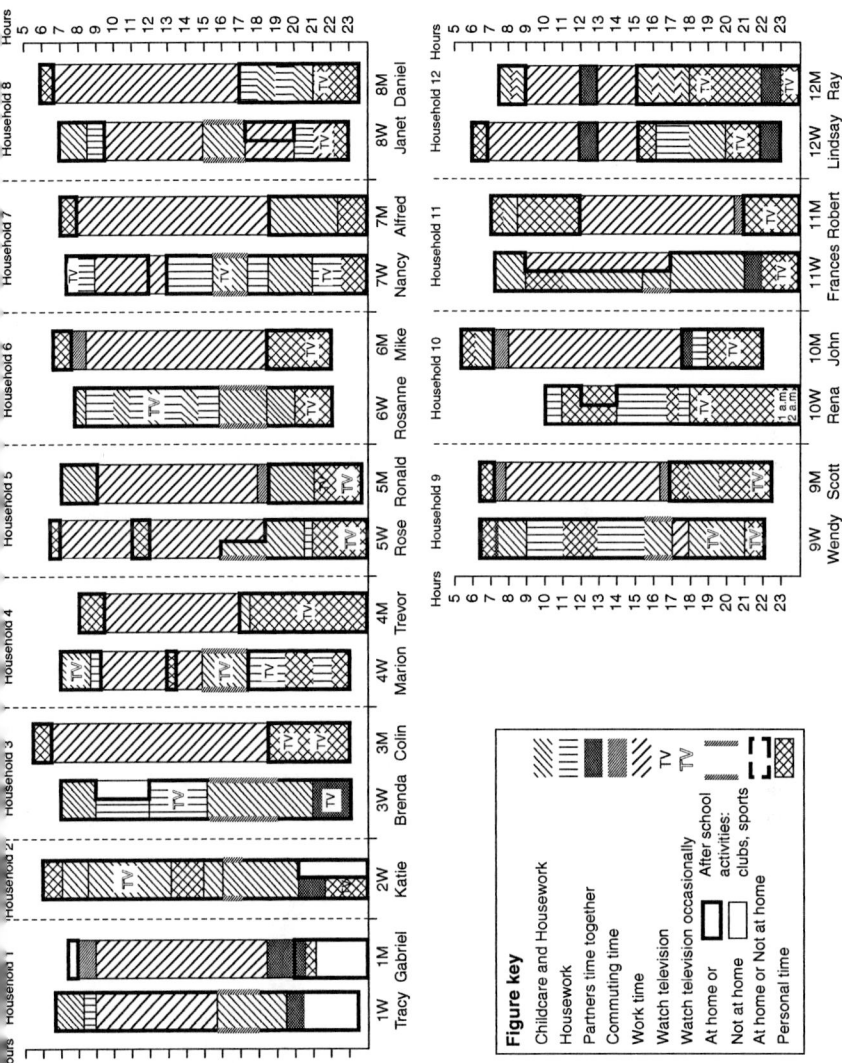

Figure 3.1a Space, activity and time of everyday life – I

64

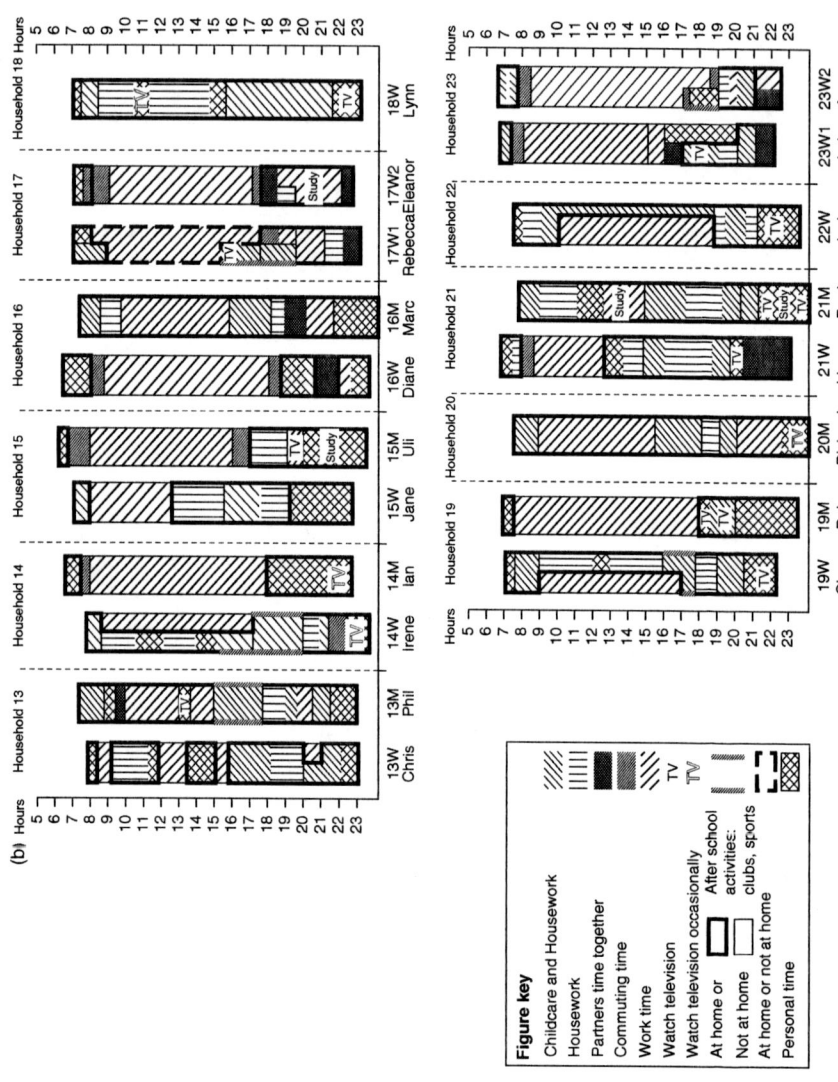

the individual with most personal time in the sample. She claimed not to spend any time on childcare (her husband did some of this). She spent some time on housework and also did a few hours a week of 'boat cleaning' to earn some money. Unlike the other women, Lynn Murray (H18) did not have a partner, but had four girls aged between five and 11 to look after. She lived on benefits and had the lowest household income of the sample. Her day combined personal time, housework (with the television on all the time) and childcare before and after school hours. For all these women, with the exception of Tracy, who had a strong professional identity, spending large amounts of time in the home was linked to devoting long hours to housework and to childcare. However, staying at home per se does not demand these activities. The gender of the person who stays at home is significant for their engagements with housework and childcare only insofar as it combines with work interests and employment positions. The routine patterns of the men who stayed at home the entire day are revealing in terms of these significances in the use of domestic time.

What did the men do? In household 16 (H16), described in Chapter 2, Marc Churchill had a routine involving childcare and housework in the mornings, working from home between 10 am and 4 pm, and also from 8 pm to 9.30 pm, after an interval between about 4 pm and 7 pm when he would again be involved in childcare and housework. In the evening he spent time with his wife and had some personal time. Richard Bartholomew (H20) undertook childcare and housework until his son went to school, working from home until the child returned home, after which he spent some time with his son and in housework activities, occasionally cooking, eating and washing up. He worked again for a couple of hours in the evening and relaxed with the television until about midnight. Fred Al-Thompson (H21) had a more broken routine than the other two men. He was involved in childcare and housework for longer hours. He had younger children (aged three and nine) and his own health required that he spend more personal time resting. His working time, spent studying, occupied him for a couple of hours a day. Marc and Fred had women partners who were in employment. Marc was in a well-negotiated role of the stay-at-home parent of three children, aged nine, 11 and 15. Fred's partner had a part-time job, since his health did not allow for him to look after the home and children on his own. Richard was a lone father of a 15-year-old boy. Compared to the women, these men were less involved in housework, while work or study occupied them relatively more than the women who also spent most of their time in the home. This is consistent with the statistical trends we saw in the last section.

Time spent in the home was very significant for most people and women spent more hours in the home than men, as shown in Figures 3.2 and 3.3. Only 16 individuals out of 42 spent their waking hours away from home for longer than or as long as they spent in the home. Of the 16, 13 were men and three were women. The fact that women spent longer hours in the home than men echoes a relatively traditional gender pattern, but variations were not simply due to gender. For instance, three men did not go out to work, and while five women also did not go out to work, five other women had their everyday routines mostly away from home (three of them were lesbians).

Thus, variations in sexualities, work and gendered patterns are significant. The visual 'busyness' of the routines of women in Figure 3.2 contrasts with the relative calm of men's routines in Figure 3.3, which were more spatially settled in either 'out of the home' or 'in the home' and were more clearly dedicated to singular activities. Women's time was more often broken into different spaces and activities throughout the day. This was the case whether the women spent the day mostly at home or had outside employment. In general, men tended to do more when women were not present.

Time is a resource, and being in charge of one's own time, and of the time of others, signifies having individual and social power. Although women generally have less power than men in heterosexual family life, inequalities tend to be less marked where both adults operate in a balanced way in their 'caring routines'. The care of children was central to the shape of household routines. Activities for adults (men or women) were generally organised around the needs of children and normally only one adult was involved. Where the man cared for children, he also did housework (but rarely the laundry – see Chapter 5), though usually no woman was present. Whenever a woman was present, it was she who undertook the childcare (Rena was an exception), although a man may have helped, particularly in the evening routines (with homework, tea, play and bedtime rituals).

These findings corroborate the argument that women and men use time differently because of their distinct life histories. Gender is significant to variations of time because women's subordinate economic and social positions result from practices of power that constrain their abilities to make decisions about the use of their time and that of others. However, in the same way that I contested the assumption of an intrinsic conservative role of technology in the home due to patriarchal

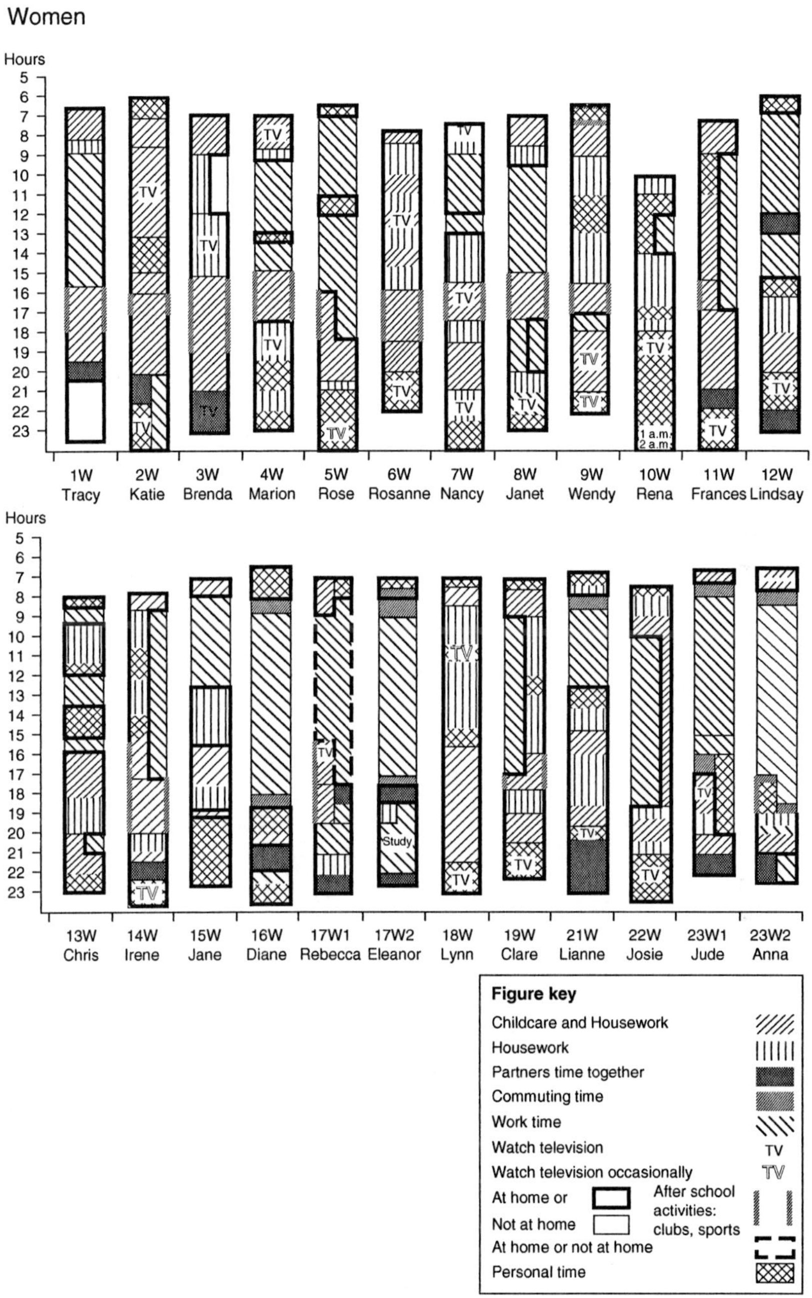

Women

Figure 3.2 Space, activity and time of everyday life – women

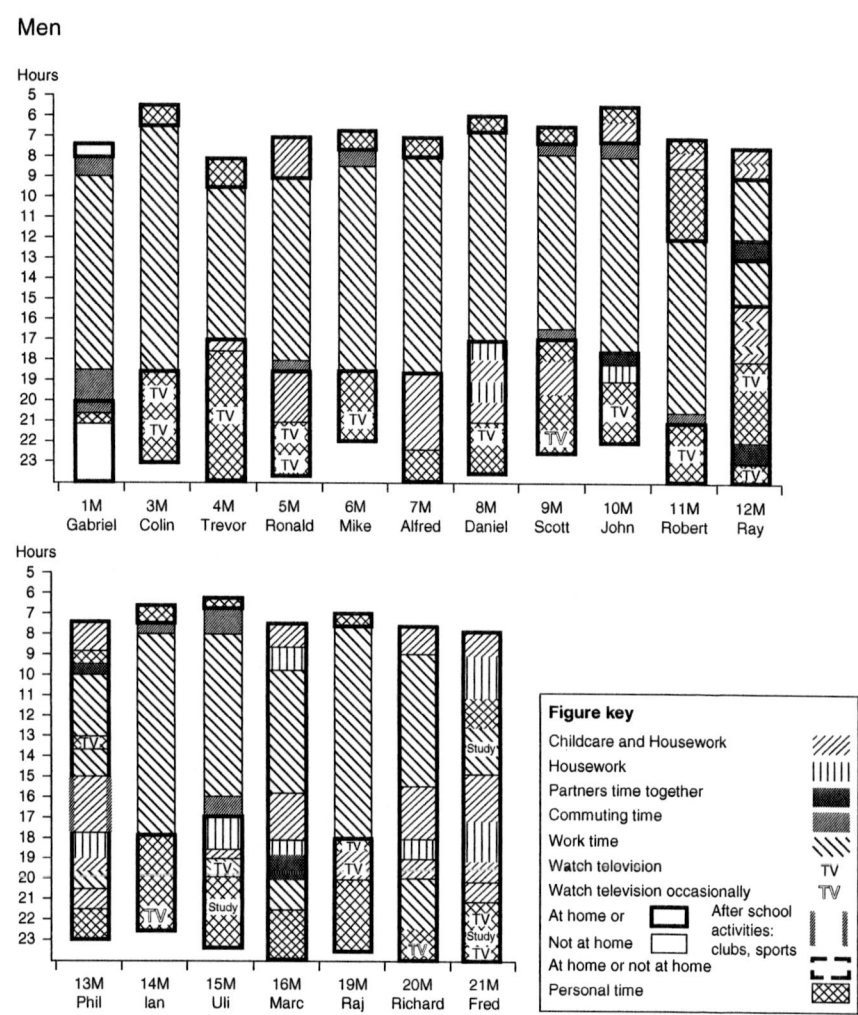

Figure 3.3 Space, activity and time of everyday life – men

forces, the opposition of largely undifferentiated categories of powerful men and powerless women needs to be qualified. As with men, female experiences are not homogeneous, despite particular forms of masculinity and femininity prevailing during different historical periods. Gendered times are constructed in relations between women and men and they also change during the life course. The patterns we see in Figures 3.1a and 3.1b were established in relation to concrete situations of the labour market and in interaction between partners. Although this interactive process does not indicate consensus about the outcomes, there was little sense of complaint or resentment about these routine patterns in the individual narratives. However, current conflicts in the home between women and men are often conflicts over time (Silva, 1999b). Some men used professional commitments as an argument against the demands of women that they took on more work in the home (see also Hochschild, 1997). This may be a conscious strategy that some of them, like Gabriel Green (H1) and Robert Gibson (H11), were not ashamed to confess:

Elizabeth: ... so you are at home by eight-ish?

Gabriel: Yeah, usually home for eight or if I choose to, I'll ... just after seven.

Elizabeth: ... do you tend to see your kids before they go to bed or not?

Gabriel: It goes through periods ... I tend, I tend to see them ... But – ... to be quite frank, I deliberately avoid getting back ... – I tend to, to not rush back.

Elizabeth: Is Tracy conscious of that?

Gabriel: I think she probably is, yeah. I certainly am.

Elizabeth: [Talking about Frances Gibson serving her husband Robert a meal at 9 pm everyday] Do you think you should do more?

Robert: Eh?

Elizabeth: Do you feel you should do more?

Robert: I don't feel guilty, put it that way!

Elizabeth: Why not?

Robert: Don't see any reason why I should! No, I think it's just, I don't know really, I suppose it's just habit.

Frances worked part-time in a bank, in a job-share scheme. Robert's work time lasted from midday until 8 pm. It is notable that in Frances' narrative of her everyday routine, serving Robert a meal at 9 pm was a very matter-of-fact activity – uncontentious, taken-for-granted, very much in the same manner Robert saw it. Frances said:

> ... Robert works every day, ... he finishes work at half past eight and gets home about nine o'clock at night so when he comes in at night, I then make a bite to eat for him, and I have a bite with him, and then we settle down to watch television at that point.

For Tracy, who worked full-time from home, Gabriel's pattern also appeared uncontroversial. She said:

> [The kids are in bed at about 7.30 pm] When they're in bed – I usually have supper with Gabriel when he's back and maybe about half past eight, quarter to nine, I will go and do another two or three hours' work if I need to.

Intersubjective negotiations appeared to be very relevant for the gendered locations of these individuals. These may have been related to more conventional patterns of gender relations, as seemed to be the case with Frances and Robert, or they may have resulted from modern reflexive choices to fit everyday life issues to life stage circumstances, which is more in line with Elisabeth Beck-Gernsheim's (1998) argument. Gabriel recounted:

> ... when they [the kids] were younger, things were quite different then because – I worked from home for two days a week, Tracy was lecturing at the college and I used the technology to move my hours around it. So I could take the kids to school and collect the kids from school, ... I'd work on a computer in the middle of the day and late in the evening and I'd still do a full day's work but I'd be looking after the kids as well. So – from that point of view, the technology enabled me to enable Tracy to, to, to do her teaching. I wouldn't have been able to do it any other way ...

Despite traditional resonances in these patterns of gendered uses of time, the most relevant observation from the patterns of time use in this sample is the presence of various new experiences of everyday routines, related to different ways of living, away from traditional Fordist models of a male breadwinner and a female homemaker. Looking at Figure 3.3, we see that the two men who spent the longest hours out of the home were Gabriel (1M) and Colin (3M), followed by other five men who also spent long hours on outside work: Mike (6M), Alfred (7M), Daniel (8M), Ian (14M) and Raj (19M). However, only four out of the 18 men, Gabriel, Colin, Mike and Ian, presented a pattern of traditional male providers who at the time did not get involved in any childcare or housework. While Gabriel was prevented from getting involved by his long commuting hours (even though being prevented felt comfortable), the other three engaged in this role with conviction. But while Colin's wife Brenda worked out of the home for a few hours some days a week, and Mike's wife Rosanne was the most traditional of the women in the sample (see the description of her in Chapter 2), the traditional full-time homemaker role did not fit Ian's wife Irene, who had a part-time job, which she combined with carrying out all the responsibilities for childcare and housework. Home for Ian, as for Collin and Mike, was an uncontested place of personal time and replenishment. For Gabriel, some tensions were present.

The involvement of men in feeding children and themselves (for example, Phil (13M), Ray (12M) and Marc (16M)) and the not-a-big-issue approach to feeding husbands in the evening, illustrated by the quotes from Tracy and Frances above, indicate that the effort in these activities was not to be remarked upon. The easiness of tasks and the ability to combine managing the home and children with working for pay was achieved by, on the one hand, the availability of technological aids and related objects and practices for everyday living and, on the other hand, by established patterns of daily routines. These set routine patterns of dealing with housework and the technical artefacts in the home usually involved a negotiated gender division of labour (in the less traditional homes), the help of older children, and also the help of other adults (see Chapters 4 and 5 for more on this). Some of the better-off households bought some form of domestic time as they paid for cleaners, babysitters and childminders. All of these helpers were women. Some poorer but also some well-off families enlisted the help of grandparents (more often grandmothers) in childcare, particularly with the school run and after-school care. This is evidence of the new social divisions between women I referred to earlier (see also McDowell, 2008).

Only four couples of individuals mentioned having the same getting up time. While the morning routines were quite separate for partners living together, with each one going about their own affairs, the evening saw the appearance of their being together and of the television in the narratives, mostly after the children went to bed. Daytime television watching figured little in the routine accounts of women and men. Six women watched daytime television. Nancy Mitchell (H7) turned the TV on first thing in the morning, when coming into the kitchen, and left it on until she went to bed, but the others had shorter set times of viewing. Only one man, Phil Webster (H13), who worked from home, timed his lunchbreak by the BBC One O'Clock News. Children in eight households watched television before going to school. The TV served adults as a kind of 'electronic nanny'. However, the TV was on for most of the day in a number of homes. During the fieldwork research period, televisions were found on at different times of the day in nine households. These included all those homes that used morning television as 'nannies'. Having the TV on did not mean that people were watching any programmes in particular. For many, it was on as 'background noise' (Katie), 'for company' (Nancy), 'so I don't feel alone' (Lynn) or, as in Richard's account: '... a lot of the times I just – I end up sitting in this state where I just literally drift over in my mind, so even if I'm watching TV a lot of the time I'm not really watching TV, it's just things drifting over in my mind.'

Evening television watching fitted this escapist mode for most people. Television was placed within a personal time of relaxation or of relating with one's partner. This was the pattern for 24 individuals. However, six women and four men did not account for a role of the TV in their everyday lives. This does not mean they did not ever watch television. Households 1, 13, 16 and 17 were generally TV-free. There were particular characteristics about these virtually TV-free households. Tracy and Gabriel (H1), both architects with a young family, caught up with work in the evenings. In household 13, Phil, an actor and writer, worked from home and was in charge of the daytime care of his four children. Chris various jobs ('lollypop lady', school playground supervisor and hairdresser) meant that she came in and out of the house at various times of the day and evening. In the evening they both shared the care of the children and had some time together. Marc (H16) worked from home and looked after the two younger children outside of their school hours. Diane worked long hours and she did not have 'hands on' involvement in the daily affairs of the home. This was the only household where a conscious decision not to own a television was taken. Diane said: 'I experimented with a TV in my bedroom. It was not healthy for the relationship. Not having a TV makes us go out. We spend evenings together, chatting. I talk on the phone to a friend. I don't want one. We don't, as a couple.' In household 17, Rebecca sometimes worked from home and Eleanor out of the home.

Evening routines changed depending on whether Rebecca's daughter was there (usually for half of the week), but time was mostly spent replenishing oneself or the relationship. In these four households, individuals were high in cultural capital, valued relationship and personal time and made conscious choices of time use in daily life. This behaviour aligns with findings in empirical research I developed with colleagues on cultural capital, where a strong correlation was found between shorter television viewing hours being connected to purposive television viewing (as opposed to watching whatever happens to be on), and social position, among those with higher education, workers in the cultural industries and in the executive and professional class (Bennett *et al.*, 2009).

Time constructions were linked to dual-time schemes of workdays and holidays. For most people, real family time happened at the weekend, mainly on Sundays. Sundays and other holidays were perceived as extraordinary because then people had control over the time. Personal needs and set routines were then less focused, and investment in the self and in relationships was centre stage. The Sunday routine may have involved going to Mass, going on family walks or engaging in particular forms of leisure, but more often it meant eating together. In a few cases this meant a proper Sunday dinner. Often the meal was just a time when people living together were available for each other. Certainly children's affairs were prominent in the narratives of everyday routines. But the analysis of time in home life, in particular the extent of 'choice' in routine patterns, appeared as a subject bound by complex explorations about the boundaries of the public and the private in people's lives. Women's work has been less and less governed by a logic of male-female exchange in the private sphere. The 'transformation of time' noted by Adkins (2009) bears heavily on the contemporary condition of womanhood, as working time and non-working time are increasingly difficult to differentiate.

Time changes and variations: summary

The temporal orders of reference for the individuals in this ethnographic study varied mainly according to gender, labour market participation and the presence of small children. Even when they were not in paid employment, paid work gave a general meaning to domestic time, due to its overall social dominance in contemporary life. This happens because when time is 'contaminated' by emotions and affections implicated in caring routines and relational practices, the distinctions between what is public and private are ambiguous, and the boundaries between them are increasingly fluid and changeable. Clearly the choices of individuals and households showed messy, fragmented and diverse patterns of everyday routines. The key setter of routine patterns for women and men living with school-age children was school hours. Secondly, it was paid-work hours, but these also varied. In making up the household routines, negotiations were crucial. The care of children was central, but the gender of the carer (still mostly female) and the ways in which caring practices were organised varied according to each household's particular arrangements.

It became apparent that although patterns of everyday life, routines and habits must be firmly established for the effective conduct of messy everyday affairs, there were no set frameworks for this. This corroborates trends pointed out in recent studies of the sociology of the family and taken up in mainstream sociology, which note that family forms are moving away from fixed or rigid notions of the 'proper' family. Even inside traditionally structured families, like some in this study, changes take place. New types and qualities of relationships have appeared in practices of 'doing' family life. Routines are of great importance in this 'doing'. They are diversely constituted and may change according to circumstances, but they do not change at random, or suddenly. Household technologies, both as commodified market relations in themselves and as enablers of the adoption of other market relations, are inserted into daily routines and contribute towards negotiations of different patterns of gendered work in the home.

Ways of 'doing gender' are relevant in this context. On the basis of the patterns described here, and comparing the Fordist and contemporary contexts of gender relations and material life, we can conclude that the carrying out of the activities of housework-as-care (material and emotional) has changed over time. The uses of time in relation to technological change in the home are evidence of this. The places (or sites) where care is exercised have moved and the boundaries between women and men, home, work and the market are not neat or clear. Shifts in the gender of houseworkers are still small, albeit not insignificant. There have been changes in the time allocated to housework activities and in the ways of accomplishing them. Changes in everyday life routines go hand-in-hand with wider socioeconomic (including technological) transformations and have a strong bearing on society and culture. The framework developed by Bourdieu helps us to understand these relations to time insofar as the experience of time is regarded as part of both the objective potentialities defined by one's material conditions of existence and the subjective expectations attached to these material conditions (Bourdieu, in Wolfreys, 2000). Shaped by the past, the present habitus embodies certain expectations, like the one that housework and childcare are mostly a female domain, which create tensions in the present. Yet, it is in the mismatch between the habitus and the chances presented by the social world that social change is created. Tensions cause change to happen and these can be identified in very ordinary activities like cooking and cleaning, as I will explore in the next two chapters to further consider the interplay of individual practices, social patterns and social changes.

4
Cooking

Practices of cooking offer a means of entering into an ordinary area of everyday life to explore the accomplishment of relationality between differently positioned individuals and between individuals and materiality in the form of the position of individuals in social space, processes, ingredients, technologies and results achieved. Methods of going about cooking may be elusive or reflexive, but they are often intersubjective and have a time dimension. While conditioned by past practices, they are patterned by current material and cultural conditions as well as by expectations.

Katie Hughes, when reflecting about her cooking practices, conveys in her account a notion of social change, evident in her ambivalent assessment about the sorts of rules to follow and the valuation of the outcomes of her actions:

> ... if I was to define the word 'cook' I'd say that you're starting from scratch and you're using your own vegetables and your own ingredients when you're making your dish. I don't know what (word) I'd use for the other, opening the freezer and putting it in the oven but I suppose it's cheat's cooking. No it's not ... it's lazy, – convenience food if ya like. No, we tend to have takeaways about twice a week, which is another bad habit we've got into. (Katie Hughes, H2)

Cooking as a technical process is the application of heat to food. Heat can be applied in different ways, either directly to what is to be cooked or to the varied utensils containing ingredients to make dishes. In addition, cooking is but one of the activities in food preparation and one of the components of the job of feeding a family. Food preparation involves a number of smaller tasks: planning meals, shopping, setting the table, preparing and cooking food, serving the food, clearing the food and dishes from the table, putting leftover food away, washing dishes or loading the dishwasher, drying dishes or unloading the dishwasher, putting dishes away, cleaning counters and stove, wiping the table and sweeping the kitchen floor (Luxton, 1980). Feeding is an equally complex set of activities surrounded by emotional and ideological notions, while also relating to bodily reproduction and to lifestyles. These have been thoroughly researched by Marjorie DeVault (1991) and Nickie Charles and Marion Kerr (1988). These comprehensive analyses do not, however, refer to the implements with which cooking is accomplished. Cookers are absent agents. Yet, cookers are central to the limits that time and money impose on what to cook, how to cook and when to cook, as well as regarding the expectations of standards of cooking. In this chapter I focus on cooking technology, like cookers, microwave ovens, fridges and freezers, and on cooking routines to consider the relevance of the resources involved in cooking practices for family relationships and how the practices identified provide means to analyse social change. To explore cooking technologies, the framework of Latourian-inspired actor-network theory (ANT) is relevant, while my reflection on cooking practices and relationality draws more directly from Bourdieu.

Cooking technologies

In the previous chapters I began a discussion about the processes by which practices are implicated in the design of machines and in uses of time related to gender. Gender practices and gendered time are included and become constituted in technological development. Three aspects about these connections can be distinguished in the feminist literature on technology (Wajcman, 1991; Cockburn, 1992; Cockburn and Ormrod, 1993; Cockburn and Fürst-Dilic, 1994; Gomez, 1994; Chabaud-Rychter, 1994, 1995; Ormrod, 1994; Berg, 1996): (1) both gender and technology are processes; (2) they are shaped, or acted out, in interaction; and (3) they are both culturally and historically contingent categories. In this section I explore the relationship of the material and the human as intermeshed in historical processes with attention to innovations relating to the cooker.

Drawing on ANT (Latour, 1988b; Callon, 1989; Akrich, 1992) and on a feminist poststructuralist perspective (cf. Scott, 1988), the developments in cooking technology can be examined considering the 'relational materiality' (Law, 1999) configured in cookers by various agents in the process of development and design, production, marketing, distribution, sales, maintenance and so on. Users transform the prescriptions of these agents when they act with the technologies, yet the technologies are themselves social agents. As I stressed in Chapter 1, the methodology offered by the ANT approach enables the reconstruction of practice (as an activity), talk and translation between actors. It also allows for an investigation of the inscription of gendered interests, politics and power.

While I share the view espoused by ANT of technology as doing, not as being, I also follow a perspective which regards gender as 'doing' (originally from West and Zimmerman, 1987). This incorporates a notion of practices through which contexts for changing gender subjectivities are captured by examining relationships between technologies and users over time, which involves seeing how gender 'appears', or how gender is performed, in cooking practices. For a number of women, gender identity is still, ambivalently or not, located through household tasks. However, the construction of gender identities, like that of technologies, relates to moving social and relational processes achieved in daily social interactions (DeVault, 1991; Butler, 1990) as part of localised dynamics (Thévenot, 2001). Activities in households and the instruments used for the performance of these activities convey certain normative gender roles. We can see in Table 4.1 that cooking has largely been a female affair. Yet, roles change over time (in the 25 years between 1975 and 2000, men's cooking time doubled and that of women nearly halved) and society increasingly offers more than one single gendered normative script. Changes in households with children have progressed more slowly and those with lower education had a greater reduction in time spent on food preparation in the period than those with higher education, possibly because for the latter there had already been considerable reductions.

To show that cooking technologies involve conceptions of gender that, though materially implicated in ordinary social relations, are constituted

Table 4.1   Time spent daily on food preparation: UK population aged between 16 and 65

Characteristics		1975 mean minutes	2000 mean minutes
Gender	Male	11	23
	Female	100	58
Education	GCSE or lower	57	45
	A Level + further education	50	37
	Higher education	45	34
Young child in household		65	60

Source: Based on Cheng *et al.*, 2007, Table 1. Data from Multinational Time Use Survey for 1975 and UK Data Archive for 2000, Office for National Statistics.

historically and have never been stable, I explore two major innovations. The first is the thermostat oven control, introduced in the 1920s and 1930s. The second, dating from around the mid-1980s, is the microwave oven. I reflect about the transformations of subjectivities – as gendered – in cooking practices in the home[1] and provide a background for considering the cooking practices of contemporary families in the following section.

Joan Scott (1988: 42–4) proposed four levels of analysis for the interpretation of how gender becomes implicated in social processes. These are: (1) 'culturally available symbols that evoke multiple (and often contradictory) representations'; (2) 'normative concepts' that fix the dominant meanings of such symbols; (3) institutional complexes in which these are articulated; and (4) ways in which the relevant symbols, norms and institutions are implicated in the construction of subjective gender identities. Scott argues that these elements could apply to analyses of how gender hierarchies are constructed, legitimated and maintained or challenged, providing methodological guidance for an interpretation of how gender appears historically in the materiality of technology (Scranton, 1995). But this framework is mostly applicable at an abstract and conceptual level.

I examine Scott's four levels of analysis, combining them with the ANT approach. I utilise Madeleine Akrich's (1992) guidelines to expose the script inscribed by those involved in the process of innovation in the technologies (designers, manufacturers and others). These scripts define a framework of action for agents who have specific tastes, competencies, motives, aspirations and morals, as well as incomes, housing and relationships to time and to other agents (Akrich, 1992). Methodologically, scripts act as ways of integrating the wider social context in analyses of technology.

Within the first two levels of Scott's analysis (1988), it can be argued that the script of the cooker refers to the feminine gender identity and to family living. The cook is traditionally identified in the script as principally a woman, servant or housewife. 'Inscriptions' about the gender of the cook have changed historically. However, the dominant meaning of the cooker, as an object for use by women in the home, has remained central to the processes of technical innovation.

To reveal the script of the technologies, I consider the cases of the cooker and the microwave oven. For each of these, I investigate firstly the operating instructions concerning their setting. I regard this as the original script. This is not the creation of a single individual. There are many layers involved in the processes of innovation and in the creation of technology scripts. Design, development, manufacture, consumer research, creative departments, marketing, distribution systems, sales and product regulations all interact, feeding into what I call the original script. My analysis is based on a wide range of instruction manuals for cookers and microwave ovens, as well as claims about what the technologies offer to users made in advertisements and sales leaflets. Secondly, I explore the instructions regarding the cooking process of particular foods and some recipes. This exploration attempts to identify the transformations of the script by the user. The recipes are those given in the instruction booklets or in cookery books relating to particular kinds of appliances.[2] Both types of instructions relate to the third and fourth levels of analyses proposed by Scott, referring to the institutional articulations of traditional gender identities and prescriptive actions. They also contain a notion of politics, regarding the changing prescription of gendered practices, and transformations of subjectivities and household moralities. I create a contextual narrative interpretation of these scripts by treating them as texts (McCracken, 1993). My analytical strategy is close to that of Anne-Jorunn Berg (1996), who uses the notion of the domestication of technology to conceptualise the process of the decoding of artefacts. The concept of the domestication of technology, originally from Roger Silverstone *et al.* (1992), refers to the consumption of technology in households and to the processes of creating a home. The symbolic aspects of technologies are central to the process of domestication and they are linked to the household 'moral economy' (I will explore moral issues of family living more centrally in Chapter 7).

Thermostat oven control: clever cooker and leisured cook

As a reminder of the immense changes that currently available cooking appliances carry in themselves, let us return to the beginning of the twentieth century, when it took a certain amount of expertise to know how much fuel would produce a required cooking temperature. A woman born in 1902 describes this process of temperature regulation:

> Once I had a bed of coals I knew that it took four quarter logs to hot it up enough to bake bread. It only took a crumpled paper and a stick or two to boil the kettle though. If you wanted to cook muffins you stoked up the fire then stuck your hand in and started counting. If you got to eight before it got too hot to stand it, it was right for muffins. Bread was six and pies were ten. (Luxton, 1980: 134–5)

This bodily-related expertise was to be incorporated in the oven-regulating thermostat, the mechanical regulation of time and temperature, a milestone in the history of the cooker and the first notable invention in cooking technology since the middle of the nineteenth century. According to Sigmund Giedion (1948), the thermostat was invented by an engineer with the American Stove Company and it was added to the company's range of products in 1915. By 1930 this device had been widely adopted by manufacturers under the American Gas Association. However, in Britain before the Second World War, thermostatic control was available only for a few of the better quality cooker models. The earliest extensive references to the thermostat oven control concerned the 'Regulo', introduced from 1923 by the New World Gas Cookers of the Birmingham-based manufacturers Radiation Ltd.[3]

Cooker production was addressed to cater to the work of the middle-class housewife, who was singled out for special attention. She had to run her home largely or entirely unaided because of the growing shortage of servants.[4] This was the type of cook conceived as the reference in the innovation process. Hers was a new professional salaried household with a typically small family. Meanwhile, the working-class housewife could afford only the smaller and cheaper cookers that abounded in the inter-War period.

The instruction manual presenting the Regulo cooker stressed the automaticity of cooking, where 'watching, or attention, on the part of the user is unnecessary' (Radiation Ltd., 1927: v). Yet, analysis of the instructions for the cooker's operation reveals contradictions between the claims that the machine 'does it all' by itself and the actual demands of the technology placed upon the cook for its proper operation.

The claim that cooking in a cooker with a fitted Regulo was merely a mechanical operation that did not require thought or judgement is contradicted in various ways. For example, the 'Instruction Chart', which gives the recommended temperature and time for cooking various foods, is accompanied by a set of 'notes' which set particular cooking requirements. The 'notes' refer to meat, poultry, cakes, bread, pies, puddings and so on. Many foods have cooking time intervals such as '30–45 minutes', '2–3 hours' or '2 hours or longer'. Thus, when addressing real usage of the cooker, the assessment of the cooking time depends on examining and assessing the cooking process, and certainly on opening the oven door, unlike what was stated in the instruction manual ('there is no need to alter the gas or open the door for examination of the food during cooking –': Radiation Ltd., 1927: v). Yet, when referring to the technical endowments of the cooker, the instructions do not acknowledge the cook's role in the assessment of the cooking process. It is assumed that this is knowledge possessed by the machine. The cook who thought it necessary to check the food while it was being cooked could indeed feel that she was being told off by the operating instructions. In the recipes and the 'Instruction Chart' (Radiation Ltd., 1927: vii), cooking rules appear full of 'but', 'if' and 'however'. They demand assessment from the cook and in many cases knowledge of some arithmetic to define proportions of time in accordance to weight, thickness and the intensity of heat application. Unrecognised checking, changing, watching and attention are crucial features of the cooking process, as well as knowledge of the properties of different cooking ingredients.

The invisibility of the cook's attentiveness constructs her as doing and thinking nothing. The claim that the technology incorporates the whole knowledge of the cooking process universalises the cooking experience. Yet, this universalising is still 'circumscribed' to all middle-class housewives who had lost their servants. The cooker requires these particular women to desire a particular kind of role. The woman colludes with the moral vision of 'the innovator' when hiding her efforts behind the always cheerful, fresh, nourishing image of proper womanhood in advertisements and feature articles on domesticity. Such images hid the fact that she had to experiment and define the rules of her cooking:

> You think that because I haven't got a maid, I must spend a lot of time cooking meals. But it's the 'Regulo' on my New World Cooker that does the work! First I set the 'Regulo ... o[n] whatever number ... Not a thing to do once [the dinner] is in the oven – ... (Advertisement in *Woman*, 25 September 1937: 75)

At a broader level of technological development of cookers, the exploration of less conventional lines of cooking was held back because it was thought to clash with the conventional female role. Just after the Second World War, a 'leading authority' on electricity for domestic use was reported to have said that 'as a matter of policy it was felt that it was unfair to a housewife to introduce her at the same time to new methods of cooking as well as to a new fuel' (PEP, 1945: 66). Although the patronising of the housewife and her supposed inability to deal with simultaneous changes in fuel and new methods of cooking are insufficient to explain the industry's choice of investment, the perception of the 'needs' of women did affect the patterns of technological innovation and the expansion of the market. The requirements of electric cooking followed along the lines prescribed by solid fuel and gas, leaving other potential advantages of electric cooking unexplored.[5] Better cookers were not affordable to the majority of families and innovations were held back. Cooking technology could theoretically have followed a different pattern, but alternatives were slow to be explored.[6] While cultivating the accomplishment of normal femininity through cooking, alternative developments to save labour time and make cooking easier were not pursued until the introduction of microwave cooking.

The microwave oven: prescriptive cooker and neutered cook

A new cooking technology emerged in the 1970s, combining developments in electricity and electronics. The microwave was introduced as a saviour of the busy housewife. Cooking demands were said to have been eradicated:

> Your husband is returning from a conference in 45 minutes. The chairman and his wife are accompanying him. But dinner is still unprepared. You would like to offer tomato soup with cream, followed by trout with new potatoes, peas and asparagus, and for dessert, an upside down cake topped with peaches and cherries. Impossible, until now! For today in that time, you can cook the meal of your choice, set the table, put the children to bed and get changed. (Sharp Microwave advertisement, *Good Housekeeping*, September 1973)

The technology employed in the microwave, the magnetron, was first developed in the 1940s at the University of Birmingham. As Cynthia Cockburn and Susan Ormrod (1993) observe, the potential for the development of microwave ovens as known today existed from that date. However, the technology was first employed in early radar equipment. After the War, further research on the magnetron developed it as a cooking technology. It was first employed in catering. The market for microwave ovens began to expand in the 1970s, increasing sharply, particularly from the late 1980s onwards. The wide diffusion of the microwave has depended in particular on continuing innovations which transformed early microwave cooking practices by combining them with conventional cooking methods, in so-called 'combination' ovens. These first appeared in the USA in 1981 (*Consumer Reports*, 1981). Technically, combination ovens operate either by the principle of convection cooking, by which a fan circulates hot air, or by use of the thermal principle relying on the natural circulation of heat in the oven. Microwaves do not apply heat to food directly. The waves penetrate into the food, making the moisture molecules move rapidly, thus producing heat. In combination cooking, the oven cycles back and forth between microwave and convection cooking. It makes up for the rigidity of the microwave technology, offering cooking by simply microwave, grill or convection, as well as in combination.

As the quality of cooking improved, prices were reduced and demand soared. In 1987, 30 per cent of British households had a microwave oven, increasing to 74 per cent in 1996 (GHS, 1998) and reaching 91 per cent in 2006, as shown in Chapter 2 (Table 2.2). Reheatable convenience foods have grown in popularity and the diffusion trends are related to changing family lifestyles. The proportion of married women in employment in Britain, for example, rose from 54 per cent in 1973 to 67 per cent in 1996 (GHS, 1998). The proportion of partnered women with dependent children in employment increased from 59 per cent in 1994/95 to 70 per cent in 2005/06 (Simon and Whiting, 2007). Households with dependent children had the highest ownership (84 per cent) of microwave ovens (*Family Spending*, 1997).

The proper operation of microwaves made instructions and recipe books highly important. Yet, like the inbuilt buttons, numbers and lights, instructions were generally rather obscure. Surveys have stated that consumers use microwave ovens mostly for defrosting and heating up food that is already prepared. In my ethnographic study, this was also the case. In 1991, in the UK, only about 25 per cent of those owning a microwave oven cooked with the appliance (Cockburn and Ormrod, 1993: 148) and in the USA, this proportion in 1990 was five per cent (*Consumer Reports*, November 1990: 733), a difference that reflects diverse cultural patterns of cooking in a broader sense.[7] Microwaves seem to have basically been major enablers of growth for the market of pre-prepared foods.

Why is there so little 'cooking' done with microwave appliances? Is there a failure of 'innovators' to communicate or a lack of skills among cooks? Or is it that parallel market developments in pre-prepared food have appropriated the technology to different needs, which equally cater for the demands of versatile lifestyles?

There are considerable limitations for proper cooking with the microwave. Microwave cooking guidelines and recipes tend to be specific to the model of the oven. The construction of 'intelligent cooking' is thus based on a dedicated machine capable of replicating results under rigid settings. The machine imposes a rigid script on the user. This is hardly the way 'real' cooking is done. As in the case of the thermostat oven control, microwave cooking does not free the cook. Minding the microwave is a required task when cooking, involving checking and assessing. The more powerful and quicker the cooker, the greater the importance of timing in the cooking process. Moreover, often the time required for cooking is not much (if at all) less than that required for cooking in a conventional oven. Yet, in the typical product literature, microwave cooking claims to abolish the cook. The cook is ultimately dissolved and unnecessary:

> Ever confused by cooking instructions? – Ever had to guess cooking time? – Found it difficult to calculate wattage differences? ... The Neural Network is trained to recognize the type and weight of food ... and then decides how long to cook it and at what power level ... (Sharp/UK, 1995)

However, the unrecognised cook, presumed to be abolished in the chain of ready-meals-from-freezer-into-microwave-to-table, retains great significance in the process of stocking the freezer and in combination cooking. Nor is heating chilled ready-meals unproblematic. Instructions and recipes for microwave cooking usually refer to a 'universal' appliance and most microwave cookbooks and convenience food instructions are written with full-sized ovens in mind. However, small ovens, with lower wattage, take about 30 per cent longer to cook the same foods. For a single baked potato this means one minute more or less, but for a casserole the difference could be 10 or 15 minutes. Users have to adjust the time to their machines.

Less apparent are the continuing requirements of the tacit knowledge (the untaught skills acquired through normal living in a particular culture which derives from one's habitus) of the cook that arise from the limitations of the technology. It does not in fact make the cook redundant, but shifts the requirements of the role. Assessment in the cooking process – ingredients, dishes, temperature, time and operations – continues to depend on the tacit skills of the cook.

The cook has to assess the cooking, but some cooking knowledge is also appropriated into the machine. The trade-off is that freedom for cooks comes at the cost of certain rigidities in the choice of raw materials in order to guarantee the replication of the cooking process without human intervention. For example, it is recommended that only Maris Piper potatoes should be used for baking jacket potatoes (Panasonic, 1990: 83). The homogeneity of the raw material ensures replication. The engineering of agricultural products is in this context the ultimate dream of the appliance manufacturers, because it enables the complete control of the cooking process with no intervention from the user, and proper universalisation. Another example of rigidity is the recommendation that only parboiled rice should be used for cooking in the microwave. For dedicated use the technology does well, but in the face of diversity and complexity it demands a skilled cook. In this context, definitions of what cooking is changes and newer scripts for the user emerge: licence is given for women to be reckless and new types of cook appear, fitting with the different practices found in households, as can be seen in contemporary cooking routines.

Cooking routines

Time involved in the preparation of food has declined. As a matter of course, contemporary homes do not have cooked breakfasts or midday meals. In my study, when asked to indicate the most useful machine in the home, only one in six chose the cooker. The microwave oven was the least useful for one-third of adults. Does this mean that cookers are invisible, or is it that people do not care about having them? How central are cookers to eating?

Irene Hays-Field (H14) chose the cooker as the machine she couldn't do without 'cos you could wash your clothes by hand but you couldn't cook in any other way, ... going out and lighting a fire' she wonders, pensively, before stating '(the most important) ... it would have to be the cooker'. But not everyone sees cookers as being that useful and, interestingly, two individuals chose the cooker as their least useful appliance. They were both in low income, ethnic minority, lone parent households. Richard Bartholomew (H20), who lived with his son (15), shared cooking with the mother of his children, who lived in a different house with their daughter (8). The mother did most of the cooking and they tended to eat meals at her place. He occasionally cooked in the evenings. Lynn Murray (H18) seemed not to eat very much at all. This may well be because there wasn't enough food for everyone. Money was very short. She said:

> it's normally just a bit of cheese on toast or a sandwich ... I'm not one for meals. [Referring to her four daughters she added:] They may eat scrambled eggs on a Friday. ... They like these things – these noodles, these pot noodles or egg and toast, ... they just nibble really a lot, they're not one for a big meal. ... I stopped doing Sunday dinners completely – I mean I do like a Sunday dinner, roast beef, Yorkshire puddings and t'gravy and that but they won't eat it ... – it's dear so I don't bother buying it.

But Lynn used the microwave oven, which had been acquired three months previously with a loan off 'the Social', daily. Lynn was ashamed of her kitchen and had been waiting for years for it to be renovated. In the future, when this happens, she hoped, she would be able to have proper meals. Until then cooking was a low priority (I will look in detail at Lynn's story in Chapter 6).

Overall, cookers were used mainly in the evenings, but sometimes for breakfast and on weekends as well. As Lynn's poverty illustrates, in some homes hardly any cooking is done. However, there wasn't much cooking done either in the well-off Bird household (H9), where it was said that because the children ate at school, cooked meals were done only on weekends. In the intermediate-class Addison household (H3), they made do with frozen pre-prepared meals and pre-packaged salads, bread and ready-made desserts. Corroborating working-class Katie, quoted at the beginning of this chapter, Brenda Addison (H3) said: 'I do cheat cooking', while working-class mother Jane Naylor (H15) emphasised that 'the kitchen, it's not my place'.

Nevertheless, a big cooker was desired in a third of homes, with three mentioning explicitly an Aga cooker and two some sort of Aga-like cooker.[8] These were wanted in either professional or intermediate-class households. Sensors for when food is ready and, in particular, self-cleaning features were highly desired design features of a 'dream cooker'. The cookers owned by interviewees covered a wide range of available makes and types. More than half were over five years old and in about one-third of cases they were less than two years old. So, cookers were appliances going through reasonable renewal. Microwave ovens of all ages were owned by nearly everyone and were of great assistance in cooking processes and in feeding practices more generally. Defrosting, warming up and reheating were the most common tasks. They were also used for snacks and individual portions. Frozen and chilled foods were integral to cooking practices. Fridges tended to be older appliances, with over half being owned for over five years and less than a quarter acquired in the last two years. Most complaints about fridges related to their small size and many households wanted a bigger one. Interestingly, information and knowledge about freezers were vague concerning their age, uses, complaints and desires. This was an appliance that was less frequently used than more active technologies like the fridge and the microwave oven. Freezers lived in the background, sustaining feeding and cooking relations without explicitly imposing on daily life. Other appliances relevant to cooking processes were the waste disposal (surprisingly mentioned in three higher-class households) and food mixers. Waste disposals are uncommon in the UK, being more consumed in the USA. Their presence in the three UK households was clearly a sign of distinction.

In the 2000 'Time Use Survey', 69 per cent of women in the UK said they liked cooking, while only three per cent said they did not do any cooking. Among men, there were 58 per cent who liked cooking, but 15 per cent who did not ever cook (Gershuny, Lader and Short, 2006). In terms of food preparation – including cooking – Shu-Li Cheng *et al.* (2007) found that between 1975 and 2000, men's participation increased by 12 minutes, while women's decreased by nearly half, reducing by 42 minutes (Table 4.1). Gender, education and the presence of small children were the main factors differentiating involvement in food preparation, within an overall trend of time reduction over the 25-year period. There was some gender transference of time spent, but overall there was a marked *elimination* of time devoted to preparing meals. In households where people had higher levels of education, there was less time spent in food preparation, a trend that has remained stable over time. Data on income is lacking, but the approximation of levels of education, the occupational structure and levels of wealth allow for an inference that it is in better-off households that less time is spent by individuals in preparing meals. While there was consistently more food preparation in homes with young children, in these households preparation time has also been reduced.

Changes in cooking practices are due to the ways in which technological innovations connect with what I call a 'technological nexus' (Silva, 1999b), which in the case of cooking involves the use of restaurants, school meals, hours of work outside the home, as well as utensils, washing up demands and so on. Convenience food is an illustration of how the technological nexus operates within patterns of social differentiation and inequalities. The convenience food that most resembles and tastes like home cooking is the most expensive. This is also the sector of the market most responsive to health warnings and consumer safety. Less tasty food that presents greater health risks is consumed mostly by the poor. Cooking enabled by the connections of new technologies like the microwave oven, freezer and dishwasher is more readily achievable by better-off households who can buy better home-care-equivalent foods for use in suitable household appliances. Buying, freezing and leaving items for household members to heat up or cook fits well with versatile, flexible lifestyles. Less well-off households have less flexibility to share with the commercial market the level of care required by their needs of feeding families. And this has some implications for how they share the jobs of cooking in terms of gender and age.

Cooking no longer requires much experience, creativity, knowledge or practical skills. Some cooking is child's play. Peter Mitchell (8), in household 7, told me how he made popcorn:

> ... well, I opened the packet and then I opened the plastic covering and then what I did was I opened the two sides and placed it, – I set it for four minutes, switched it on – and then I ... it started popping and you had to listen and it stopped popping, and when it stopped popping you'd take it out and open the two sides so it was – all stopped and – ... you just pour it and ... eat it!

Yet, cooking may involve things like experience, creativity, knowledge or practical skills, as well as be a form of 'gift giving' activity. Alternatively, it can be anything else not involving all of these. The status of the activity of meal preparation is varied. When done for display it carries explicit notions of distinction and gendered divisions become more salient: more men cook for display than as a matter of routine. Feeding children carries status, even though what is implied in this feeding has changed the significance of the emotional aspects of cooking. Currently, the social and symbolic value of food provisioning is created with the shopping and the addition of small personal touches. As Katie (H2) said, a typical meal is 'something from the freezer to the oven with a bit of fresh veg thrown in to make me feel not guilty'.

The main obstacles to seeking commercial solutions are cost, which is where inequalities are most notable. Money matters in this increasingly commodified world. The unequal distribution of income has a direct impact upon the provisioning of food and, as Alan Warde (1997) remarks, it is possible to examine which class a person belongs to by examining their food tastes and eating habits. A working-class diet remains distinct from a middle-class diet. Class distinctions can also be found in the cooking practices in the home and in the ways in which these combine with market provisions, as part of the technological nexus.

Although the sharing of domestic cooking with the market provision is more limited for poorer households, it is nevertheless prominently done. Lynn's claims that her children nibbled a lot, ate pot noodles and disliked her cooking, which she gave up doing because it was costly and got wasted, is indicative of the penetration of the market into family feeding in poor households. Consequences for health and lifestyles were not a concern for Lynn. Conversely, in one of the wealthiest households of the sample (H16), Diane Churchill was not a cook in her house. She said: 'I quite like cooking, actually, but I just don't have time, ... I don't want the kids to live on junk food and equally you can't afford ... to get (every-day) ready cooked meals in Marks and Spencer ... too expensive.'[9] Thus, she managed the daily food preparation undertaken by her husband and son, sorting out the weekly menus and shopping lists. Eating together freshly prepared meals was important to the family ethos. Acquisitive power and health concerns in cases such as these go together, irrespective of the time available for cooking. Lynn had lots of time available, Diane had none. Lynn had no kitchen with suitable equipment to cook in, whereas Diane did. The resources of these women, including Diane's high cultural capital, her knowledge of diet implications and her ability to draw from her husband's and son's time for cooking, are markedly different. These resources also relate to dimensions of care (Chapter 6).

Time budgets can never reveal the emotional and caring dimensions of labour. Charles and Kerr (1988) show that individuals are aware of the symbolic and comforting role of family meals. DeVault (1991) offers a detailed appreciation of the caring aspects of food provisioning, considering the social and emotional coordination. These aspects carry considerable weight in gendered patterns of cooking practices: women consistently do more, even those feeding families in the twenty-first century. Does this show a continuing and pervasive conventional division of labour? No, it does not, and there is abundant evidence of changing cooking practices in contemporary family life. However, the link between women and cooking confirms the special place of feeding in the domestic division of labour and the patterns are rather nuanced.

Professional and managerial men participated more often in food preparation activities in the study by Charles and Kerr (1988) and men did more when the woman was in full-time employment (see also Warde, 1997). Alan Warde and Lydia Martens (2000: 96–9) found that men contributed less in households with children between the ages of five and 15, but that their contribution was greatest when children were younger. Warde and Martens distinguished between the central tasks of planning, cooking and serving the main meal, and the ancillary task of table setting, clearing up and washing dishes, finding that men only ever gave a greater contribution to the ancillary tasks. It remains the case, they argue, that the norm is that food is a woman's responsibility. University degree qualification increased male contribution, as did full-time dual-earner households, but again these affected mainly the ancillary tasks. Significantly, nearly 20 per cent of the men in households investigated by Warde and Martens had cooked the last main meal alone. Men do cook nowadays, changing the prevailing view stated by Liz Stanley (1995), based on 1940s Mass Observation data, that 'women cook and men eat'.

Following broad statistical trends, *men who cook* were found prominently in my study,[10] but their everyday practices appeared subtly disputed in spontaneous appraisals of the divisions of labour. The disputes addressed either standards or the amount of effort and level of involvement with the task. In all of the households where men cooked, people had paid employment, although Chris Webster (H13) had short working hours but a very fragmented routine, while Ray Wells (H12) worked from home (see Figure 3.1b).

Chris (H13) said that cooking was 'shared 50/50'. 'I never say to Phil "oh would you do tea, I don't feel like doing it will you do it?" It just seems that one of us around say like ... what shall I make for tea, or in the morning Phil might say "I'm gonna make a lasagne tonight" or whatever and that's how ... how it goes.' Phil, her husband, chose the cooker as the most important appliance in the home (for Chris it was the washing machine). He presented his cooking role differently from Chris's narrative, saying he was the one who cooked the most in the house, doing it everyday.

Jane Naylor (H15) did not like cooking ('I'll cook if I really have to', she said) and Uli did most of it. Their oldest son, Will (13) also did basic stuff, regularly cooking for the family on Tuesdays, when his dad was out. He used the microwave 'a lot' for hot breakfasts and sandwiches. While Uli was the one taking the role of the main cook, he expressed a sense of being harried in the tone in which he said: 'I don't have a lot of time to cook.'

Ray Wells (H12) was cooking a large amount of burgers to feed his large family when I first came to their house. Lindsay, his wife, said about his cooking: 'Ray makes convenience food ... but he's very good, because he's around a lot ye know, he'll do tea ... or he'll go shopping.' Ray said: 'I would start the tea most nights ... There is always seven people for tea here every night, quite often we have more ... so if there's something in the freezer I'll just use that. ... I would have thought Lindsay decides, ... gives the ideas.'

The households where *either men or women* took the cook role with no sense of trouble were the homes of professionals with a highly egalitarian gender ethos. Marc and Diane Churchill (H16), as I remarked in Chapter 2, had strong egalitarian gender principles, which they applied to home management. Marc cooked three days a week, and Greg, the son, two days. Diane did the menus, instructing them to cook simple things. Greg learned from his parents and was also studying cooking at school. The microwave oven was used very little because they did not have 'many ready-cooked meals'. In household 5, Rose Chambers currently cooked, but it used to be Ronald. 'Why the change?', I asked. She replied: 'I guess 'cos I enjoy doing it, he didn't – he used to cook, well ..., he's – he can't just throw things together, he has to look at a recipe book whereas I just, I quite enjoy throwing things together so I'm much relaxed, he gets very uptight ...'

In some cases *a well established and more conventional gender division of labour* prevailed, like in households 14, 9, 6, 7 and 8. For example, Irene Hays-Field (H14) did the cooking but Ian might do it on Saturdays if she was at work. Ian said: 'I can get fish fingers out of the freezer and ...' The children's perception was that it was always mum who cooked. Similarly, Wendy Bird (H9) said: 'Scott cooks sort of main courses and I do all the other bits and pieces.' Asked 'who cooks', she said: 'Scott cooks, we both cook.' Scott agreed that he cooked at weekends and that he was 'basically a savoury man, I can't stand doing puds, like all good chefs'. Their children said 'mum cooks the most'. In Janet Seaman's case (H8), there was a professional involvement. She was professionally qualified in cooking and was demanding of her appliances. Daniel cooked too, 'stews and proper dinners', but the children didn't like his cooking: '[what] daddy's does is rubbish.' These were cases where cooking appears the most to resonate with femininity, with women generally taking over the task and their men doing the occasional support or cooking for display.

Cases *where only women are cooks* were rare. They either lived as lone mothers, in lesbian households or, in heterosexual homes, resenting the assigned role. For example, Clare MacDonald (H19) said 'there aren't different days when people do different cooking: it's *me*! Unless I'm ill! And then we'll get beans on toast!', concluding with reference to Raj's cooking. A different tone was expressed by Eleanor Hill (H17), who said that probably Rebecca Turner did more of the cooking: 'I'm not really interested ... but I actually do a lot more cooking now than I ever did in the past', acknowledging that she was 'learning with Rebecca'.

Men who do not cook may be more commonly found among the working class, according to statistical evidence, but clearly are not class restricted. They are, however, a category larger than that of women who do not cook. Some dabbing with foodstuffs is often a part of everyday life for everyone. In household 3, Colin Addison said 'I don't cook, no':

Elizabeth: Not at all?
Colin: No. I do – well I cook ..., when I say no ..., I don't mind doing a breakfast, a fry up. I can –

Elizabeth: When do you do that?
Colin: Well, on a Sunday morning.
Elizabeth: Actually, I heard that you cook wonderful breakfasts.
Colin: Ah!
Elizabeth: The children seem to like it!
Colin: That'll be Sunday, YEAH. It all goes in for me.
Elizabeth: What do you cook for this breakfast?
Colin: Bacon, sausage, egg, mushrooms, beans.
Elizabeth: Do you enjoy doing it?
Colin: Well no, but I'm t'first up so I do it, it's easier to do it, I always
wake up ...

Gabriel Green (H1) said he never cooked because Tracy was a good cook, adding that 'I can't achieve her standards, you see!'. I asked him: 'If she lowered her standards would you do more?', and he replied: 'I don't, I don't like cooking. I find it tedious and futile.' Tracy, his wife, said: 'Gabriel rarely cooks. He can do omelettes and pancakes. If I'm ill or go to London for the day he is capable of doing it. I'd rather he spent his time with the children.'

Children's involvement in cooking involved using microwave ovens a lot. Children learned from seeing their mothers, and very occasionally fathers, cooking. But mothers may be reluctant to let their children get involved, as in the case of Marion Lakin (H4). I asked her if she taught her boys how to cook and she replied:

yeah, yeah, I mean they'll watch and they'll help but there's some-times – too many cooks spoil the broth ... kitchens are dangerous places ... you can't have like four people standing round here ye know ...

If cooking and feeding the family still remain predominantly the affairs of women, these accounts show that transformations in the feeding responsibilities of women and men are, however, notable. Three main trends running alongside each other can be noted: (1) the accomplishment of normal femininity through cooking; (2) the licence for women not to cook or to be reckless; and (3) the appearance of new cooks. These transformations are derived from cooking practices of individuals in households, yet they are somewhat dissonant from the relational materiality emanating from prevailing gender conceptions of the technological innovators, which reaffirm women's traditional role as cooks.

Resources for cooking as practice and care

In contrast to the 1930s cook of the thermostat-controlled oven, the contemporary cook is no longer homogeneous, and cooks are no longer primarily caregiver females. Newer categories of cooks are identified in explorations of domestic life and in advertisements related to lifestyles and family practices. The gender identities constitutive of microwave technology are fragmented and include men (young and old, who cook and/or help), children (girls and boys), women who do not like cooking (or do not know how to cook) and women who are good cooks. Advertisements currently invite women to feel less uneasy about not expressing love 'properly' (Warde, 1997: 132). In addition, food has come to have value outside and beyond family situations (Warde and Martens, 2000). Shopping occupies more time in everyday life, with the obligation to express care through domestic labour becoming less relevant. These trends confirm Silver's (1987) argument, which I introduced in the previous chapter, that in the long run women's employment would increase demand for formally provided services, one of these being cooking.

While in technological developments cooking and caring appear connected in the generalised acceptance of cookers as naturally belonging to women and feminine domesticity, it is notable that in the commercial world of restaurants and catering, cooks are generally men, and they are then called 'chefs', a term appropriated by men in everyday life when emphasising the specialness of their cooking contributions (see Swinbank, 2002). Yet, the connections between cooking and female caring have not been stable. In both the innovations of the 1920s and the 1980s, it has been claimed that intelligent and capable appliances grant freedom to 'the housewife', though the images used to represent the intelligence and capability of cookers in the two periods differ. The early images of maids or of magic to ease the cooking process are no longer invoked. The 'Perfect Maid' (Hotpoint, 1930s) has been turned into 'The Neural Network' (Sharp, 1990s). The 'real' labour of women (maids and housewives) was reconfigured with additional brain support. In these two historical moments, cooking effort takes on distinctive appearances, bringing about newer categories of cooks and carers in the contemporary home.

However, one way by which feeding families has encompassed new cooks involves the greater penetration of the market provision of food via the invitation to stock the freezer with cooked meals – whether home-made by mother or bought – to be heated when needed by those who need to eat (see also Shove and Southerton, 2000). It is implied that these are men and children, the new 'cooks' of microwave technology. The invisibility of feeding the family is enhanced, with food always ready in the freezer. Patterns of cooking-as-care have been redefined.

It is significant that it is in the simplest cooking operations that the microwave has its best 'brain power'. However, it cannot 'think' well when doing the more complex and differentiated jobs of cooking, which tend to still be a predominantly female area. Such technological scripts reveal different moralities of objects and gender. They reveal how household technologies are constructed in relation to certain dispositions and practices in society, which relate to the normative expectations of gendered everyday living in households. The ways in which microwave ovens have been 'domesticated' have clear connections with the moral economies of the household. Yet, as we have seen, gendered expectations have changed, with complex implications for family living.

The technological innovations regarding cookers have not been responsible for all the changes in cooking practices and changes in related gender roles. Yet, a focus on innovation in cookers shows that a good quality cooker can make cooking and related activities easier, but also that machines make limited difference in isolation from social contexts. For example, since the 1990s, not only has the cook of the microwave technology changed, but current representations of thermostat oven control in advertisements now also emphasise the technology rather than the cook. The cook has most noticeably disappeared from the picture, in contrast with the ever-present housewife-woman in the pictures and words of the 1930s Regulo advertisements (Silva, 1999b). However, the cook has not disappeared. The main user of the cooker remains a woman, although other users have appeared. But, with increasing 'intelligence' in the machine, used as part of the technological nexus, it has also become possible to address neutered cooks, for instance, 'busy professionals' or 'overworked parents'. This enlarged category of cooks reflects a reshaping of gender boundaries. The kitchen is no longer an exclusive realm of women, just as the world out of the kitchen is no longer exclusively a place for men. Actually this transformation of gender-related properties appears to have some basis in class divisions, with the kitchen becoming a place of sociability and distinction, in particular for better-off families.

A recent wave of lifestyle television programmes on food and cookery in the UK promoting celebrity chefs have attracted a large audience. Yet, in analysing the influences of such programmes on cooking practices, Tim

Lang and Martin Caraher (2001) indicate that the effect has been minimal. Sandie Randall (2002, following Wood, 2000) similarly finds that people like to watch these programmes but do little cooking inspired by them. Perhaps the influence of the programmes, like those of the expanding market of cookery books, relates more directly to combating time scarcity through indulging in time consumption, as Joanne Hollows elegantly notes (2008: 129). It is not the juggling of domestic life – or feeding or eating – that is of concern here, but the enjoyment of life. Escaping from the problems of the 'time squeeze' with the 'intensity' of alternative experiences of domestic life in order to fully indulge in domesticity (baking, creating fancy dishes, making grand presentation of foodstuffs) is a sign of distinction. This is more pleasurable for the middle classes, who are more likely to have the option of buying the labour of other poorer individuals (usually a woman) to perform the less gratifying tasks of washing up, chopping food stuff, clearing and tidying up. Significantly, this division of labour in cooking tends to reflect the hierarchy when men cook for display at dinner parties: women often take on the supporting role.

These patterns refer specifically to the roles of gender and power in the processes that make certain kinds of technologies and access to media enjoyment available to particular sections of society. It reveals a close, but complex and not straightforward, historical affinity between how gender appears in activities in the home and the changing conceptions of gender and class that are embedded in cooking technologies and cooking practices. This reflects both Bourdieu's view of a stratified social world, which affects home living, and Latour's take on the ways in which material technologies are agents in social action. The actual operations of technologies in the home can only be clearly seen by looking through both these lenses.

I have argued that there is a connection between the construction of the identity of carers (predominantly in this case of women as cooks) and of the technological development of cookers. The ANT approach reveals that the scripts for the operation of the technologies express the normative gendered expectations in society. However, the actual usage of technologies, as expressed in recipes and everyday cooking practices, tend to depart significantly from such norms. The 'scriptwriters' often express more a wish than an actuality, whether in the prescription of not opening a thermostat oven door, in contradiction to the needs of assessing the cooking process, or in the microwaving of a perfect jacket potato or tailor-made joint of meat. Significantly, scripts only make sense in the context of a technological nexus and range of skills. Hierarchies of gender and power are implicit in the processes of technical innovation, but the operation of the technologies is patterned by everyday demands negotiated according to localised, hierarchical, uneven and complex demands which shape collective practices. I examine the operation of these patterns again in the next chapter, in relation to cleaning activities and related technologies.

5
Cleaning

Cleaning has not inspired much scholarly curiosity and we know little about contemporary domestic cleaning practices.[1] Historians have paid more attention to the subject, revealing past practices and processes related to the design of houses and home furnishings (Rybczynski, 1986), technological innovation (Giedion, 1948; Forty, 1975), social and economic effects of technological consumption (Cowan, 1983; Mohun, 1999; Parr, 1999) and the jobs of workers and servants (Malcolmson, 1986; Sutherland, 1981). Cleaning bodies has also been a matter of historical concern focused on hygienic prescriptions (Hoy, 1995). Historical changes in notions of comfort, cleanliness and convenience related to bodies and clothes have likewise been investigated as part of broad technological systems and practices of cleaning (Shove, 2003). Apparatuses and ways of cleaning bodies, as well as cleaning houses, have constantly revealed demarcated and reproduced inequalities through the practices of different social groups.

Cleaning is a very large subject involving the making of ideals of order (Douglas, 1966), political processes and ethical judgements as much as entrepreneurial imperatives and technological opportunities (Strasser, 1982; Swasy, 1993; Parr, 1999). It has chiefly involved large amounts of the mundane work of women, much of it invisible in the home and also to scholars (Gronow and Warde, 2001), much of it not distinguished from other related activities in official statistics, as shown in Table 5.1. For example, in cooking activities, a certain amount of cleaning of foods, utensils and machines are involved; tidying is often done in the process of cleaning; moreover, laundering involves sorting clothes, adding cleaning agents, drying, folding and/or ironing, separating and storing. All of these are carried out by women and men, but women do the most overall.

This chapter concentrates on the washing of clothes and dishes, relating these to developments and uses of washing machines and dishwashers.[2]

Table 5.1 Time spent on housework activities per day (UK, 2005)

Activities	Women mean minutes	Men mean minutes
Cooking/washing up	54	27
Cleaning/tidying up	47	13
Washing clothes	18	4

Source: Based on Gershuny, Lader and Short, 2006, Table 5.3, p. 50

I investigate changes in these practices and their connections with technological innovations, focusing on effort, time use and social relations. Changes in families and households are explored when looking at the literature on the effects of technological changes in cleaning. These address some of the paradigmatic views about household technologies which affect both industrial policies and feminist politics. The technologies and activities of cleaning clothes and dishes, and the narrative of practices of two households – the Rock and Churchill families – complement this reflection, highlighting and challenging issues in current debates.

Practices of laundry and dishwashing have been affected by the manufacture of suitably designed appliances, the design of the home, income levels and standards of living, the composition and age of the family, whether women have outside employment and for how long, the presence of children and their ages, and the use of paid or unpaid help. I begin with an exploration of the historical assembly of these elements in laundry practices.

Laundry practices

A thread that I have been pursuing in this book concerns the interaction between the ways in which the doing of things and the instruments available to do these things are mutually shaped within particular socio-cultural contexts. In line with the outline I presented in Chapter 1, this means that the ingredients involved in laundering – the fabric, the detergents, the conditioners and the machines – are not innocent materials, but that they affect practices and contribute to constructing ideas of what washing is and how it is, or can be, done; who might do it, when, where and so on.

As it used to be

In around 1900, the family laundry took about seven hours a week (Lebergott, 1993: 112). It was done by a woman with the aid of buckets, a tub, a wash boiler, a washboard, a scrubbing brush and a cake of hard soap. She first carried water in from the well or the communal tap, pump or stand-pipe (as shown in Chapter 2, it took some time for plumbing to reach large numbers of houses). She then filled a wash boiler, carried gallons of boiling water to the wash tub and begun to scrub the dirt off the clothes. Once the clothes were clean, she had to carry away several pails of dirty water, fill the pails with clean water and get on with the next load. When the wash loads were done, she would wring the clothes out by hand or lift them straight up, heavy with water, and put them through a separate wringer with rollers. Wringers could be adjusted according to the bulkiness of the clothes. The woman would then hang the clothes up to dry (which would be done indoors if the weather was wet). Most clothes still *needed* ironing and airing. Doing the laundry was a relatively simple task, but it required a considerable expenditure of energy. Ironing, however, was a skilled job (Mohun, 1996).

In the early 1900s, electric motors began to be attached to washing machines, to move the dolly (a special stick inserted into the tub to stir and agitate the clothes) and also to operate the wringer. By 1916 the electric motor was sealed to the bottom of the tub. As noted in relation to cooking technology (see Chapter 4), improvements in cleaning technology also increased as 'good' domestic help became more and more difficult to acquire by middle- and upper-class women. Up to the early 1950s, job advertisements for domestic servants made reference to the availability of 'electric equipment' for housework or 'modern conveniences' in the employers' houses in order to attract help (*The Lady*, 14 November 1946: 440). A decade later, these claims had disappeared, since it was by then taken for granted that the homes advertising for domestic help would possess these technologies.

In the 1920s and 1930s, the range of themes in advertisements for washing machines was more diverse than after the Second World War. Although the main foci were on automaticity, safety and health, issues of space and storage, the specialisation of the machine and scientific explanations of the washing process were also used. The automaticity of laundering was linked to effort, time and skill: 'the washing-day troubles are solved', 'it takes half time to do the laundry', 'look at Mummy's washing doing itself' (Bendix and Servis advertisements, 1956 in *Good Housekeeping*). The need for a woman to operate the machine, which at the time required considerable involvement, was kept hidden in the claim that the burdens of laundering were eliminated by the technology. I do not think many women believed this, but this sort of advertising inaugurated a phase in which laundry jobs came to be seen to be as difficult as 'pressing a button', as Gabriel Green (H1) commented when talking about laundering practices at the beginning of the twenty-first century.

Safety and health claims were related to the clothes, the woman and the environment of the home: 'no damage to material, buttons, fasteners', 'it requires the very least exertion', 'you stay upright while using the machine', 'keeps water from the floor' and 'no steamy surroundings, or scorched hands or arduous labour' (see the list of advertisements in Silva, 1997a). The issue of effort was recurrent. Scientific claims explained how the dirt was worked out in the washing process, giving a rational explanation as to why the machine achieved better results than when laundry was done by hand. The industry's efforts combined the prevailing practice of 'small articles of clothing washed at home' with the use of commercial laundries for the bulk of the laundry, preserving the privacy and flexibility of more personal laundering. To increase the market, customs needed to be changed and marketing took this into account: 'there are still many people who imagine that clothes wear better when washed by hand. Probably, however, the reverse is the true case' (see Silva, 1997a).

Alternatives to household-based laundry included commercial laundries, communal laundries in public washhouses and hired washerwomen who worked either in or out of the home. In Britain, the use of bag-washes was most common well into the late 1950s (and was still used in the 1980s in launderettes): large items of clothing like bed sheets and towels were sent to commercial laundry services (which offered a collection and delivery service), while smaller washing machines were then employed for personal clothing. The efficiency of commercial laundries was clear in the great number of families who used the service for heavy washing: 80 per cent of the population in the poorer London districts used wash bags (PEP, 1945: 61). They had insufficient space for drying clothes and no access to technology. The vitality of commercial laundering rested on the underdevelopment of home appliances. The sector gradually floundered once laundry, with its wide industrialisation from the 1950s onwards (Malcolmson, 1986), moved increasingly into the house, meaning that clothes came to be washed predominantly in homes and mainly by women. It is these sorts of developments that are captured in Cowan's (1983) book title *More Work for Mother*.

In the second half of the twentieth century, two major themes in laundry innovation were increased automaticity, in terms of machines being able to do the whole range of laundering jobs, and efficiency, measured in terms of time and cleanliness. Washing machines were able to adjust the washing time, water temperature and number of rinses required. But manufacturers' competition, as expressed in advertising, still emphasised a dream: 'just imagine, one machine to cope with all ... – put in dirty clothes, take out clean, *fully* dry clothes ...' The dream refers to the creation of the washer and tumble dryer unit, which was perfected in the mid-1980s but had not gained a reputation of efficiency even by the twenty-first century. Many users' practices make up for technological shortcomings, as the accounts below illustrate.

The reasons for why laundry with washing machines in private households became the predominant form of laundering in the second half of the twentieth century has been explained along three lines. One is that it responds to a normal capitalist logic of producing more merchandising and of privatising the work of women (Strasser, 1982). The second explanation is that laundering at home has been a response to a social desire to preserve family autonomy (Cowan, 1983). The third argument stresses the ability to complement various household activities such as washing clothes with minding children or doing other household tasks (Malcolmson, 1986), and the exertion of greater control over personal and family hygiene (Leto, 1988). Time and money savings compared to commercial laundry also count. Actually these explanations combine in the practices of laundering nowadays.

As it is now

> Well ... what would I do if I didn't have the washing machine! I don't know, I'd have to go to the launderette ... I couldn't hand wash that amount of clothes ... I mean I love my dishwasher but, I mean ... if you didn't have it you'd wash up, wouldn't yer? – but you couldn't do all the week's (laundry) washing could yer? (Brenda Addison, H3)

In my ethnographic study, the washing machine was chosen as the most useful machine in the home by all women, except two, for whom the top choice was the cooker. Lindsay Wells (H12) stated: 'I couldn't cope without the washing machine – I'd go into panic stations if I do without the washer.' For men, washing machines were neither the most important nor the least important, except for Colin, Brenda Addison's husband, who chose it as the least useful, his most useful being the television! Yet, in the Addison household, with two adults and four children (one being a baby), 29 loads of laundry were done every week, the highest for all families. Very intensive laundry practices were also found in the Seaman (H8), Rock (H10) and Naylor (H15) households, where over 20 loads a week were laundered. Six households said that they laundered between eight and 14 loads, 11 did between four and seven, and only in three homes (H17, H22 and H24) were less than three loads done weekly. Size of families and age of children were important variables explaining differences in practices, but the ways in which families engaged with cleaning practices in general appear to be affected by other concerns, as the detailed analysis of the cases of the Rock (H10) and Churchill (H16) households in the section below illustrate.

In only four homes was the washing machine more than ten years old, and there was a correspondence between greater intensity of use and newer machines, although the length of partnership formation is also related to the age of household equipment. Tumble dryers, together with microwave ovens, were chosen as the least useful appliances in the home, but while microwaves were used frequently, tumble dryers were very rarely used in most homes. Only in H24 was it used whenever clothes were washed, although in H12 it was used most days. In H4 only underwear was tumble dried, because it consists of 'small bits and pieces harder to sort out on the line', while in H16 the opposite practice of not tumble drying underwear (because 'they are delicate and may shrink') and confining it to the larger items prevailed. The avoidance of tumble drying clothes was mainly related to concerns of economy and the environment, together with a feel for freshness in clothes, and the fact that some clothes got hard or shrunk. In the summer, clothes lines were used outdoors, but in the winter, the main dryer was the heating radiators.

Often children, and most men, expressed a magical idea that clothes got clean effortlessly. 'Put clothes in the basket and they get cleaned', exemplified Henry (H24) ironically. Systems varied. In the more gender-equal households, tasks were divided up somewhat 'naturally'. Lianne Al-Thompson (H21) set the machine on and Fred emptied it and hung up the clothes; similarly, Jude James (H23) loaded and Anna Cox 'unloaded and pegged'. However, only five men were involved in weekly laundering. Richard Bartholomew (H20) was a lone father and voiced the most matter-of-fact statement about laundry: 'I make sure that stuff keeps turning over rather than getting a big pile of things ... I mean, it's not as if you're hitting them on a stone on a stream or anything is it?! You just chuck it in with some soap powder.' Marc Churchill lived in a serious gender share arrangement, detailed in the case study below. Phil Webster, Fred Al-Thompson and Ronald Chambers shared on what was perceived as an 'equal' basis, although Ronald said 'there is some tension at some time ... I feel a bit resentful that she's not helping'.

The lack of effort expressed by Richard or by the magical ideas that clothes wash themselves contrasts with the design of the dream washing machine for Irene Hays-Field (H14), which she describes as:

> ... one that never lets any clothes (colours) run into each other, – and one that would get all those stains that never, ever come off, marks off children's sweatshirts and things that – and – I suppose a quiet one, and – one that automatically emptied itself – yeah! And automatically matched up all those odd pairs of socks that never match up ...

Women were invested in laundering much more than men, and clearly washing clothes appeared as a much more definitively female domain than cooking. Men's avoidance strategies were also obvious. Robert Gibson (H11) said: 'I might put the clothes in but I'd – I think I've – I wouldn't risk setting it up without asking Frances 'cos I'm not *au fait* with how to use, or what setting to put stuff on.' Janet Seaman (H8) said: 'Daniel only did it twice in 17 years. On both times he ruined it. On purpose, I think, so he wouldn't have to do it again.' Less strongly, Daniel said he didn't 'usually do the laundry 'cos I've got some clothes mixed up ... I don't do it now ... She doesn't let me use it, I don't know enough about it so I don't try to get to know things like that'. Marion Lakin (H4) said: 'Trevor can actually do all these things, quite happily ... but he knows and he knows that I know that I won't be satisfied with the way he does it!' Yet, Trevor said Marion did not know how to do the washing 'because all the whites come out pink'. Often it is in the comments about particular housework activities that criticisms and comments emerge, revealing the everyday domestic dilemmas involved in obligations and the give and take of living together (see Chapter 7; see also Kaufmann, 1998).

Laundering was highly prominent in some women's lives. Janet (H8) had it ingrained into her routine: 'The first thing I do, I open the curtains all round the house, curtains open – and then washing. I usually do that – 'cos I – I don't have time to do four loads of washing before I go to work so I usually do three loads at night and then set one going in the morning.' Jane Naylor (H15) followed a similar pattern: 'I usually put a wash in before we go to bed and then in the morning I hang that out while the kids are having their breakfast and – put another load in before I go to work and when I get home I take that load out and get that hung out and then I do the ironing ...' Even when routines varied, the workload was significant in most homes. Lindsay Wells (H12), talking about lots of her time being dedicated to 'putting washing away', said: 'once it's dried ... I have to fold everything up and put it in different piles for different people so they take it up to their bedrooms and put it away...' Concerning the laundry load, she remarked: 'We always have a towel each and we don't use it again so there's at least six towels a day.' Chris Webster (H13) reflected: 'How often we wash in one week? Everyday, sometimes twice a day ... Today, that's been used three times already. Bath sheets for the kids, to wrap them around. They tend to use the same one. Bed clothes once a week, duvet cover, (every) two weeks ... Bedding for the kids ... not all at once, we rotate every week.'

Laundry activities have changed considerably both because of technological improvements and because the demands for washing have changed. Ruth Schwartz Cowan's (1983) *More Work for Mother* thesis was sparked by the observation in laundry practices that as machines reduced the burden, standards of cleanliness were raised, making women work more. But Elizabeth Shove (2003: 123) contests this view, arguing that rather than escalating standards, it is more useful to explain laundry practices as the result of 'situationally specific complexes of belief, practice and technology'. I agree that as a system of technologies for housework, washing machines have made jobs easier, and although there is no imposition to launder, and indeed some homes do little of it, in others machines are constantly running. Personal styles of laundering practices have to do with resources relating to social positions as well as with socio-technical complexes which make the capabilities of particular appliances available.

The first washing machines required water to be put in manually and pumped out; they also did only parts of the whole laundering process, demanding close surveillance and considerable labour to start and stop the various phases of washing. However, by the mid-1960s, the whole basic laundry operation had been automated. Laundry time was reduced from seven hours a week in 1900 (Lebergott, 1993) to just over two hours a week in 2005 (see Table 5.1 above). This is accounted for by a shift from the old scrubbing board to the automatic washing machine and dryer, but changes in clothing materials and what is washable have also been of great significance, even though more washing is done. A major impact has been made by the use of disposable nappies. Stanley Lebergott (1993) estimated that a 1900 American housewife washed 40,000 nappies for her four children (4,420 nappies per child per year), but by 1990 there were only two children in the average family and 85 per cent of nappy changes used disposables. In 2001–2 in the UK, disposables accounted for 96 per cent of the market, and one child would use 3,796 of them, over the average two and a half years she or he would be in nappies (Environment Agency, 2008: 2–3, 12).

Innovation patterns in laundry technology by the 1970s emphasised 'quiet machines', 'a variety of programmes for a variety of needs', the smooth working of the machines, the elimination of the 'need to iron clothes' and the fitting of automatic door locks. A new theme emerged, which I call 'cultural washing'. This refers to the suitability of different water temperatures for proper washing, part of a debate which reflects scientific controversies on the merits and dangers of hot or cold washing. A body of chemical scientific knowledge informed the best results, but there were also cultural preferences. The Germans preferred to run their washing machines with cold water, whereas the British believed that only hot water would wash well (Badden-Fuller and Stopford, 1992). According to the Good Housekeeping Institute, 'programmes on the British automatics are better geared to our kind of washing than those on Continental models' because of 'the types of detergent powders we are used to' (*Good Housekeeping*, October 1973). All this implies that product design features needed to take into account the different cultural contexts of national markets, which have grown in importance in the flexible manufacturing markets of post-Fordist times.

Since the 1980s, key innovations have related to design features such as 'energy savings' and 'half load buttons'. Newer features have appeared, such as the 'shower wash system', claiming to save water, heating, sewage and clothes. Advertisements have addressed a twin concern for economy and the environment which came to be embodied in the appliances in the form of greater intelligence: machines capable of weighing loads and calculating the water and electricity needed. Increased automaticity also meant less work for operators (women). Innovations in the laundering system addressed what I call 'making it lean', involving the rationalisation of product features and consumer choices in soaps, symbols and programmes. For example, in the mid-1990s there were 50 different laundry powders competing in the UK market, and Procter & Gamble announced the implementation of a 'great soap simplification', reducing its range to three kinds of detergents (*The Guardian*, International, 15 August 1996: 11). Similarly, it was found that 85 different washing symbols were in use for 19 different functions, with some manufacturers combining colour coding and letters in their own ways (*Good Housekeeping*, November 1996). Also, most automatic washing machines had up to 16 wash programmes, but only two were commonly used. The reduction by manufacturers of the range of offers came in response to newer requirements concerned with environmental savings.

I now turn to developments in dishwashing, which were somewhat different, before comparing both cleaning processes and related practices.

Dishwashing practices

Dishwashing has obviously varied according to the availability of piped water (cold and hot), the quality of crockery, washing up tools (plastic buckets, detergents, drainers, taps, mixers, dishwashing machines, waste disposals), space, the number of meals eaten daily in the household, the kinds of food eaten and the ways in which meals are served. To some extent, who does the job, for instance their personal resources and expertise, has also affected dishwashing practices.

Indoor plumbing provided the major improvement to dishwashing methods since the 1800s. Electricity was also crucial. Although already invented by 1851, the basic concept of the dishwasher could not be implemented, since electricity was not yet available and the hand-powered machine was not efficient (Giedion, 1948). Electric dishwashers appeared in the USA in the 1930s, but as late as 1945 few households had one. However, in 2007, 62 per cent of American housing units owned a dishwasher (AHS, 2008), a situation dramatically different from that in the UK. In 1981 only four per cent of UK households had dishwashers: since then, the dishwasher ownership has grown faster than that of any other 'white' consumer durable, rising to 37 per cent in 2006 (see Table 2.2). UK households with dishwashers are generally affluent and usually include children (GHS, 1993), which accounts for a larger than normal uptake among the participants in my ethnographic study, where only seven households did not own one (see Table A2). In the USA in 2001, 83 per cent of households with an income above US$75,000 owned a dishwasher, compared to 40 per cent among those earning between US$15,000 and US$30,000 (EIA, 2004). Market research has found that there is great market potential to be expected from the increasing proportion of women working full-time outside the home, who are less willing to spend much time in the kitchen. Dishwashers are seen by this group as labour-saving appliances (Mintel, 1995; Key Notes, 1996; EIA, 2004) and my findings also emphasise their further role in ordering the kitchen.

Stories about the burdens and benefits of different dishwashing methods show variations in relation to class and gender dynamics over time. In the inter-War period, dishwashing was more commonly seen as drudgery in middle-class homes, the employment of servants acting as the only possible escape route, as illustrated by the story of a (obviously white) young woman who stopped working as a secretary upon marriage but whose husband did not earn enough in England to employ servants:

> There is always the washing-up to be done. Dishes and forks and things stuffed in a bowl. ... four times a day, and twenty-eight times a week, and that's a hundred and twelve times a month – I'd hate to think of it by the year (...) I'm going somewhere abroad. What about the Cape, sweetheart? (...) You have a lady who is as black as your hat and says 'yes, Baas' to do it for you. (...) We won't go on washing-up and hating it. (*Woman's Own*, 5 November 1932: 124–5)

In the story, the solution to the washing up problem was to move to South Africa. While appearing as an exaggeration of the dishwashing burden, the couple's relationship is portrayed around washing up. At one point they are given a dishwasher by an old auntie but the machine does not work well. Only in the deployment of another woman's labour is a solution to be found. And this was no longer possible for professional households in the England of the 1930s.

Creative solutions to the burden of washing dishes appeared but did not flourish. References to alternative commercial services would only demand that 'the housemistress will put her used dishes in a container which will be collected by the dish-washing organisation and brought back with everything clean' (*The Lady*, 18 July 1946). Yet, the question of who was to do the washing up remained. 'How often do you face this problem?' (*Housewife*, February 1954). The words caption a picture of a cross-faced husband and wife standing in front of a sink filled with piled-up dirty crockery. Apart from dishwashers, other tools and appliances emphasised the freedom of time and from negotiations over who did the job: 'Dad never minds washing up since Mother bought a HAPPYMAID plastic cushioned Dish Drainer. What a time saver!' (*Housewife*, October 1957: 119). Progress was slow and by the 1960s, the disadvantages of adopting a dishwasher were still considerable. There was a chaotic assortment of widths and depths, some had heaters but some did not, some needed to be plumbed in, some were required to be fixed on hoses and taps (losing time) and the layout was incompatible with most shapes and sizes of bowls, mugs and dishes. In addition, not all machines were fully automatic: 'you had to stand by to turn taps on and off at the right moment' (*Which?*, December 1968). Moreover, detergents were hard to find: 'Most suburban shops don't even seem to have heard of a dishwasher, let alone powder to put in it' (*Which?*, December 1968) and it was still expensive to own crockery in large quantities (Corley, 1966). But the time-saving aspects of dishwashers were emphasised as innovation progressed and, as with laundry, washing dishes was gradually transformed into 'button pressing', with disputes over who did the dishes becoming an outdated concern, according to advertisers:

> 'What's washing-up mummy?' 'Just an unpleasant memory ... for a family with a ... ROLLS-COLSTON Dishwasher.' (*Housewife*, June 1962: 89)

> Housewives with Colstons have an extra hour each day to enjoy life – ... Husbands with Colstons never get handed a tea towel. Children don't need to find excuses. (*Housewife*, March 1965: 15)

Together with highlighting the freedom from drudgery and from stresses in relationships, advertising in the 1960s also emphasised hygiene. In a Consumers' Association comparison of washing up methods, it was found that the deployment of 'expensive' dishwashers achieved the best hygiene, but they recommended that 'if you are careful about washing up, you will wash very nearly as hygienically' (*Which?*, June 1969). By the early 1990s, hygiene was no longer a concern. Economies of cost and time became stronger issues. The weekly cost of handwashing using electrically heated water was £0.65 compared to £0.50 in a dishwasher economy programme; the daily time saving of dishwashers was 25 minutes (*Which?*, November 1991). Unfriendly high noise levels were tackled with new sound isolation material, machines became larger, with removable racks, and economy programmes were included. For the upper market segments, dishwashers were transformed from luxuries into necessities. However, handwashing continued and various other practices coexisted. In 2009, discourses stressed economies of water consumption as an environmental issue. According to government sources, washing up accounted for ten per cent of a household's daily water needs, which could be reduced to two per cent with a dishwasher (Friends of the Earth, 2009). Advertisements stressed economising water usage, as 'Earth's most previous resource', by offering efficiency with shorter cycles, less water and reduced energy usage (Panasonic, 'Weekend', *The Guardian*, 21 March 2009).

Despite this considerable technological innovation and also the larger than average uptake of dishwashers among the families in my study, washing up is still an important feature in how relationships are made and negotiated. Less of a female domain than laundry, it still occupies women significantly, revealing many aspects of gendered and intergenerational ways of relating. In the better-off households, high quality appliances required less maintenance and accomplished better quality jobs. In the families where partners had a more equal and balanced relationship, 'natural' processes of division of labour transpired by which 'if one cooks the other washes up' (H17) or 'if you leave it to drain somebody comes and puts it away later on' (H21). Men tended to be involved in loading the machine and washing up some of the big pans and containers that would not fit in, although some, like Colin Addison (H3), would never touch anything relating to kitchen jobs. Every home did some handwashing of dishes. Only when children were involved did I have detailed accounts of the washing up practices. Eleven-year-old Daniela Addison (H3) participated in a rota devised by her mother involving the three children, and described her job as follows: 'I put all the plates and things in the dishwasher, if there's anything big you wash it up, wipe down all the worktops and put all things, all the empty packets and everything we put in the bin and – that's it.' Yet, most children did not touch the dishwasher, nor did they do any washing up. The dishwasher served the home as an organiser for tidying up the kitchen, simplifying the work and ordering the space and the relations involved. Chris (H13) gave her reason for acquiring one: 'I couldn't stand the washing up as we got more kids.' Regarding washing up, Lindsay (H12), a non-owner of a dishwasher, said: 'oh, there's usually arguments over that!' She said that her oldest son, Geoff (20), 'when he was younger he had to spend every night, as soon as we had tea, he washed all the pots, cleaned the kitchen, and now – it's either I do it, he does it [pointing to her husband Ray] or we fight with the kids to do it but nobody's very happy to do it.' The detailed cases of the Churchill and the Rock families below further elaborate on these issues.

Like other housework activities, dishwashing practices relate to historical trends but, as I mentioned above, also to patterns of cultural cleanliness. I offer a personal illustration from a visit to a friend in northern England in 1980. I observed his washing up after dinner. He scraped the plates and bowls, emptied out the remaining glasses and cups, and piled them up inside a bowl placed inside the sink, which he filled up with very hot tap water, adding a good amount of liquid detergent. He then wiped the crockery with a plastic brush, dipped things into the water inside the bowl and put them, still covered with detergent foam, to drain onto the dishrack. Surprised, I enquired whether he was not going to rinse, and he replied that it was not necessary because detergents in Britain were of very good quality and there was no harm to health. After all, that was how he had lived for 30 years(!), he said. To account for the pervasiveness of this style, if one observes current advertisements for washing up liquid on British television or washing up scenes in soap operas, no rinsing of crockery, cutlery, pots or pans is seen. Most often the drying of utensils with a tea towel makes up for the lack of rinsing, taking the foam and excess detergent away. The dishwashing process is simplified through the elimination of the rinsing, but for those mindful of detergent residues, drying becomes compulsory and also creates more laundry of tea towels. A survey by the Consumers' Association in the UK found that just over a third of respondents rinsed dishes after washing and recommended that '[a]lthough it is unlikely that there is any health risk from eating off unrinsed plates, it would seem sensible, for gastronomic reasons, to rinse plates – to remove any residual detergent or perfume' (*Which?*, July 1991: 369). Housework practices are a cultural issue and washing up practices are therefore diversely shaped by different cultures. Technologies are additional resources operating in this culture.

Cleaning practices compared

In the early 1930s a survey showed that the average number of hours per week spent on washing clothes was six hours, while nearly eight hours went spent on washing up (PEP, 1945: Table 1). Laundry work was more strenuous than washing up, although it took less time. The comparison of the historical developments of laundry and dishwashing relates to technical innovations implemented in cultural contexts. Whereas laundry changed dramatically throughout the twentieth century and is now almost fully mechanised, dishwashing has been much slower to change. Why?

Let us firstly look at various aspects concerning the laundry process:

1. Laundry was a very heavy and strenuous task when fabrics were heavier, soaps were less efficient, hot water was difficult to obtain, and suitable and efficient instruments to aid in the activity were lacking.
2. Laundry was an activity separable from other housework tasks. It took place on a particular day and generally allowed for a choice of when it was to be performed.
3. The quality of handwashed laundry was generally inferior to that done with washing machines, once the technology had been improved to a certain level.

4. The development of related laundry technology reinforced the process of innovation: clothing materials, detergents and the wide use of garment washing instructions have facilitated the adoption and diffusion of appliances.

In comparison with laundry, the key aspects of dishwashing are as follows:

1. The activity has not been so heavy. In the rare past situations when families ate four meals a day at home, dishwashing required considerable effort, although dishwashing has been regarded as a somewhat 'sociable' activity.
2. The washing up job is mostly located after meal times, and it can be mixed into other tasks, unlike the non-mechanised laundering of clothes.
3. The quality issue of dishwashing has been irrelevant in most cases, with cultural (mainly class and ethnic) concerns with washing up practices that may add to effort, like rinsing, not playing a particularly important role in British households adopting mechanisation.
4. Mechanised dishwashing has not evolved into a 'technological nexus' as fast as mechanised laundry has. Special dishwashing detergents have only recently become widely available (though they are not found as easily as laundry soap), many kitchen utensils and tools are not dishwasher-proof and their shapes and sizes do not fit well inside dishwashers.

Why has dishwashing mechanisation been limited and what have been the strategies for 'eliminating' dishwashing from housework?:

1. Better 'sink technology' has been developed instead of better technology to replace 'sink work'. This involves tools and chemicals. For instance, a techno-scientific defence of the sink claims that certain grease eliminators are good for the hands of those who wash up.
2. Involving the labour of others by regarding washing dishes as a shared activity. This is an area in which divisions of labour are commonly implemented, with women involving partners and children.
3. The reduction of the amount of washing up via changes in cooking technology has involved the use of different types of fuel, performance of cookers, quality of utensils, ready-made or pre-cooked meals, the use of takeaway foods, TV dinners, school and workplace meals and restaurants, all of which are common in contemporary Britain.

Investments in washing clothes and in washing dishes by manufacturers and industrial policy makers have been markedly different. These relate to particular practices embedded in gender, generation and class patterns.

I now wish to explore how family life relates to technological consumption by contrasting two sets of washing practices which are inextricably linked to technological developments, but which largely escape any technological determination concerning cleaning methods. Here cleaning is placed within other domestic practices, revealing the interconnectedness of ways of being, social positions and the relevance of cultural capital.

Contrasting family washing practices

In this last section I consider the cases of the Rock (H10) and Churchill (H16) households to reflect on social and personal connections related to practices of washing and consider policies and politics for social change. The households present an equal uptake of the types of technologies in the home yet contrasting practices in specific choices of artefacts, maintenance and use, which express diverging practices of cleaning. While Rena Rock, who had a job outside the home for only a few hours, misused and overused her cleaning technology and always felt chased by housework, Diane Churchill, working full-time out of the home, managed her machines and the roles of family members involved in cleaning in a highly efficient way.

The Rock family

Rena, 44, was white, married to John, 50, a manager for an offshore company. They had lived in a small village in East Anglia since getting married over 20 years earlier. They had three sons: Alan (17), Geoff (14) and Patrick (12). The household net annual income was £40,000. Rena earned about a quarter of this, the lowest contribution to household income among the women in my study (except Rosanne, who did not have a job). We met Rena earlier in this book. Her earnings derived from cleaning boats two mornings a week during the summer. This paid for her annual holiday abroad with her mother and friends.

Her dad was 'in the railways' at the time she was born, in London, and her mother was 'in the offices' at Vauxhall (General Motors). She had a younger brother, living three hours away. She was in daily contact with her parents, who lived 'around the corner'. A female cousin, who was the same age as her, also visited every week. Rena hated exams, had dropped out of school before finishing it, and had had a series of jobs as an office junior, before marrying John. She did not drive because she hated exams and never took a driving test. She was 'never one for babies', but had three sons. She 'freaked out' with her first baby, was depressed with the second pregnancy and was depressed again with the third. She always found it difficult to do the housework and machines were brought in to help. The family was a very early user of the microwave oven and dishwasher. Rena found it difficult to cook and said she was overweight because she ate 'a packet of crisps and something junky' just because she found it difficult to get up and cook.

The house was a 300-year-old cottage converted from a rectory. It had been in a state of continuous improvement for the 19 years they had lived in it. The house looked 'on the go', with nothing much finished or 'in its proper place'. People were always doing things to the house and garden. Yet, things were not in working order and unfinished jobs were left around. The interior felt dark and cluttered. The fridge in the kitchen had been broken for two years. They were using the 'spare' one in the utility room, about seven metres away from the kitchen, through the dining room. The big chest freezer did not work properly, but it was important to the management of the home since they did one big monthly supermarket shopping, freezing a lot. 'I really don't know what's in there', said Rena. The dishwasher was said to be noisy but 'everything goes in there' and it was used three times a day. She said she 'hate[d] washing up' and was on her fourth dishwasher in 12 years. The washing machine was also used two to three times a day, and usually wore out every couple of years. Rena ironed every kind of garment and '(could) never catch up'.

Rena did not go into the boys' bedrooms. They each had their own bedroom and were supposed to look after them:

> if their bedrooms were tidy I'd go in there and I'd do them [the beds] but if they're not, I mean – the quilts, and everything that they've got up there is clean, they just haven't made their beds. I think Geoff's been sleeping in the sleeping bag and I think Patrick, I think he's got one on as well but I don't make them ye know...
>
> Bath towels? The boys will have a bath, they'll get a clean towel, take it upstairs and leave it in their bedroom and I don't see it for months. Alan had a pile of them, towels, in his bedroom ...

Seventeen-year-old Alan did his own laundry because, he said, 'I was just fed up with my clothes getting ruined!'. He said he complained to his mother about his jeans getting faded, and she replied 'you do your own washing'. His practice was of doing it 'just before I need to wear it', and recounted that he had 'big piles to do in my bedroom at the moment'.

When I first met Rena, she was sitting on a sofa in the living room, hand-making the hem of a heavy curtain. John and the youngest son were sweaty and muddy, digging the garden to build a wall. Remarking on the difficulty of Rena's job, I asked if she had a sewing machine. She said the sewing would 'not look right' if done with the machine. I thought of the balance between the effort she was displaying and that of her husband: both intensively labouring for their home. Their narratives of everyday life stressed industriousness and coping strategies in the face of adversities. The ordinary 'mess' displayed in the house was felt to be an adverse environment that in itself made the management of the home a difficult task, but a necessary one.

In reality, however, Rena did not appear to do very much. She normally got out of bed at around 10 am. She carried out housework for an average of three to four hours daily, including cooking tea in the evening. She watched a lot of television. John, when not working abroad (every other week), got up by 6 am, prepared sandwiches for the boys in the morning and woke them up before leaving for work. He tidied up after the evening tea. He was in bed by 10 pm, while Rena usually went to bed after 1 am (see Chapter 3, Figure 3.1a).

When John was at home he did a lot of maintenance and building work to the house. He also helped with the chores:

> I iron clothes, I cook food, I do the washing up if it is beginning to backlog up. I change most of the beds ... I can't remember the last time Rena changed the beds ...

The Churchill family

Diane Churchill was born in 1956 in a small town in the south of England and settled in London at the age of 18 to go to university. Her mother was Hungarian and a primary teacher and her father was an Anglican vicar. She had twin brothers a couple of years younger than her and an older sister. She had been in continuous full-time employment since the age of 23, when she got married to Marc, whom she had met at university. Marc was an academic, a Reader, working as a researcher and consultant in information management. They had three children, Greg (15), Hannah (11) and Alice (9). Three years after her last child was born, Diane obtained a Master's Degree. At the time of our meetings she worked as a Literacy Consultant for the Department of Education, a very demanding job which she found exciting.

We met the Churchills in previous chapters. As described in Chapter 2, they lived in a semi-detached, comfortable, Victorian house, which had designated spaces for different sorts of relationships. There was a room for Diane and Marc to lounge in the evenings, a second room where Marc worked, five bedrooms and a tastefully decorated kitchen/dining room fitted with upmarket appliances, a wooden dining table and chairs, giving a feeling of a nice place to live in.

Their net annual household income was over £80,000, one of the highest in my study, and Diane contributed 40 per cent of it. She worked long hours away from home. Marc worked mostly from home. Whereas Diane's housework consisted of supervising and organising the distribution of tasks between her husband and children, Marc was in charge of childcare and housework, including the laundry. He got up half an hour earlier than Diane, made Diane's sandwiches to take to work and organised the children's breakfast. They usually went to bed at the same time (see Figure 3.1b). Weekday cooking was done three times a week by Marc and twice by Greg. They employed a cleaner for three hours a week. Only smart clothes were ironed and only the youngest child did not do any ironing. Diane said:

> We don't iron, we fold. The ironing board is upstairs in my brother's bedroom. Marc goes there and does his shirts. I, in the summer, I iron my own things, my silky dresses and blouses. The kids' I do a bit sometimes.

Laundry was organised with individuals putting clothes to wash in designated baskets for whites or coloureds. There was also a bag for 'gentle wash'. Diane said that until the previous year the laundry had been done by the au pair, but:

> because Marc was at home a lot and Greg was old enough to cover some ... we decided that the three main jobs I used to do, look after the kids, get the evening meal and do the washing, ... we decided that Marc and the kids would take over ... and that's how it came that the kids really do quite a lot of that now.

Greg cleared up after dinner and loaded the dishwasher every day, while Hannah unloaded it, putting things away in their places. Greg did the laundry of the bedding every Wednesday; he stripped the beds, he made his own bed and his father made the other ones. Marc usually did the other sorts of laundry. He and Diane decided jointly not to use the tumble dryer much and it was used for jeans, towels and bed sheets, but not for sweatshirts, underwear or any fine clothes, because these 'shrink'. The rest dried on a line or on a rack in the bathroom. Hannah and Alice were in charge of the job of unloading the tumble dryer and sorting out the dried clothes, folding them and putting them on the bed of their respective owners to be placed in wardrobes or chest of drawers.

Diane and Marc had a deeply egalitarian ethos for managing daily life. When they started living together, Diane's twin brothers lived with them, the four of them jointly buying the house in the mid-1980s. Later on, Diane and Marc bought the brothers' share. After one of the twins got married, the other twin, Kirk, carried on living with them in a 'bedsit' on the top floor. He did his own laundry, worked shifts and was rarely at home. He had occasional meals with the family. The Churchills did not own a television. Kirk had a TV and he used to invite the three children to watch the Friday night comedies on it while he was on his night shift work. Diane and Marc always went out for a meal and cinema or theatre on Fridays. Weekly shopping was done on the Internet with Tesco Direct, despite a Tesco store being very near the house. Their appliances were of high quality, the brands chosen because of their quality reputation, and they required little maintenance.

The Churchill and the Rock households have similarities and differences. Both homes had a good level of income. Although Rena and John earned half of what Diane and Marc did – and this made a difference – life in an East Anglia village was cheaper than in north London. Each couple had three children, the Rock children being a couple of years older than the Churchill children. In Diane and Marc's home there was no television, but apart from this, the household technology intake was as equally affluent as at Rena and John's. Both households had all sorts of modern technologies. Yet background and lifestyles differed considerably and marked their interactions with technologies in ways that challenge certain theories.

Comparing practices and challenging theories

Cleaning practices reveal inequalities between social groups. As a process of creating order, cleaning carries ethical judgements and is in many ways undistinguished from other activities. The cleaning practices of Rena and Diane connect the management of their homes with how they 'did' family, how they were as people and how they were positioned in the socio-cultural space. These portray some homologies consistent with a theory of cultural capital in Bourdieusian terms. Cultural capital theory predicts a homology between social position, consumption and practices. The comparison between Rena and Diane refines some of the debates on the relationships between social and technological developments. Five themes stand out:

(1) *Saving time or using time.* The literature on technologies in the home associates their adoption and use with strategies for saving domestic (mainly women's) time. Some families, like the Churchills, consciously bought high quality appliances because performance and durability were valued highly by them. However, in other cases, as illustrated by the Rocks, saving time was of no particular concern. Or rather, Rena appeared (albeit not consciously) to want things to use up her time, to need to feel overwhelmed by demands on her time. The television was a case in point, because she watched it a lot 'to pass the time', as a way of using her time. While she felt crushed by housework demands, time was also used up with having machines not in working order, spending time complaining about them not working, adapting to new styles of doing things and in managing maintenance requirements. Conversely, Diane and Marc explicitly used the technologies in the home to save time. Their use of Internet supermarket shopping partially responded to their lack of time. Their investment in machines of higher quality explicitly aimed to avoid maintenance problems regarded as a waste of time.

(2) *Sustaining or changing roles.* In Chapter 3, I discussed the relevance of this theme in the literature on household technologies. Rena sustained a role that seemed incongruous with most women's lives in Britain. Her housekeeping affairs were not sufficiently demanding to keep her occupied. She created more work through loads of piles of laundry (bath towels were washed daily), not using the tumble dryer, ironing everything (or feeling the need to do it) and using the dishwasher up to three times a day. In some other households, however, the technology offered the flexibility to create new roles (gendered or professional). In the Churchill household, computing technology enabled Marc to work from home, while household appliances enabled him, and his family, to deal with chores. Diane wanted her full-time out-of-the-home job and Marc was happy to

work from home. Personal choices, with flexible roles, were possible *as part of* technological availability.

(3) *Using the capabilities of machines or ignoring them.* Studies of household technologies stress that the point of having machines in the home derives from their ability to enhance activities and to facilitate the accomplishment of tasks. The Rocks' machines were often out of order, despite the family's dependence on them. They did not function properly or they were neglected, remaining unrepaired for long periods. The function of the fridge in the kitchen was wasted as, for two years, the fridge in the utility room, across the dining room, was the one in use. This was a time-consuming and perverse use of the technology. Perhaps Rena's preference for doing the hem of the curtains by hand, a very time-consuming activity, could also be a case of creating time-consuming jobs, disregarding the ability of machines to assist. Of course, quality could also have been an issue, but she clearly felt the need to occupy herself and had a predisposition to ignore the capability of machines. In the Churchill household, machines were chosen for make, performance, quality and they were used – not overused – and looked after.

(4) *Using technologies to organise or to disorganise.* When machines are left occupying space and not performing their functions, they tend to disorganise activities. This happened with the Rocks' machines due to overuse, persistent breakdowns and infrequent repairs. Everyday life was frequently disorganised by the ill-operation of the home technologies. This 'disorganisation' was a characteristic of normality. At the Churchills, Diane organised two laundry baskets, for whites and coloureds, to make the laundry process easier. Marc put the laundry in, but the girls collected items from the tumble dryer, folding and placing clothes on each person's bed to be put away by the owner. Dishwashing practices were similarly organised. The family worked as a team, with rotas and responsibilities, flexibly, but orderly, managed by Diane. Individuals recognised their part in jobs and no conflict was apparent, unlike at the Rocks', where both John and Alan felt pressure and little praise for their roles with dishwashing and laundering.

(5) *Meanings of attachment and means of functionality.* Rena and John kept in the home what was not working. Rena was also a collector of old pottery and china miniatures. The dining room's central feature was a very old iron range, which they thought was about 200 years old. It was not in use. Their home's origins were in the

sixteenth century as a rectory. They liked the oldness of their environment. It seemed to give them a feeling of being rooted. Attachment to these sorts of aesthetics impinged upon the personal: they made sacrifices for it. At Diane and Marc's house, there were a few pieces of antique decoration – a vase, a picture – but they had wooden floors and sleek chairs, and a feeling of functionality and efficient home management practices prevailed. Here form was subordinated to function, with a different order of the aesthetic as integral to the personal.

Different attachments to home, relationships and the self emerge in these two cases. These build on ideas I introduced in Chapter 2. The 'personal' – with all that it entails, like choices, constraints, negotiations, relationships and so on – has an important role in the interaction between individuals and technologies: in how they are deployed and in the effects they have for lives and relationships. Yet, the personal is also patterned. The stories of Rena and Diane are inserted in class position (educational level, types of occupation), territorial location and in relation to their attached senses of propriety which inform their lifestyles. Gender and generational conflicts about what is proper to do and to be done to are part of this context (I will develop this theme further in Chapter 7).

The stories illustrate how the same sets of artefacts – cookers, fridges, dishwashers, microwave ovens, computers, television, the Internet and so on – can have very different forms of insertion in the homes, and how they can be implicated in completely different ways in people's lives. It is possible to have different practices of home life when similar technologies are consumed in the home.

Although there are often no causal relations between levels of wealth (or income), practices of intimacies and consumption patterns of technologies, there are relevant connections between them, in particular if one adopts a concept of class that involves a complex set of variables including the notion of cultural capital. One of the forms of cultural capital is 'embodied cultural capital' (Bourdieu, 1984). In these two stories there are significant relationships between bodies and machines. In a simplified manner, Rena's body and the machines in her home appeared overused and not properly maintained. Diane actively looked after her body in terms of eating healthy foods (cooked from fresh produce) and also cared for the orderly operation of her well-maintained state-of-the-art machines. Of course, these may not always go together, but in these cases the habitus and embodied cultural capital were clearly homologous, as tends to happen in wider social patterns (see Bennett *et al.*, 2009: Chapter 9). A 'relational resource' is a way of describing how cultural capital can yield advantages within relations of partnership. These are not simply relations between people, as relations between individuals and technologies in the home are also relational resources, which have implications for positions in social space. The advantages attached to Diane's lifestyle and personhood appear greater, since while she deployed these aids for everyday life, she used these resources as instruments for securing her and her family's position in social space.

6
Consuming and Caring

Practices of consumption and care are inextricably bound up with issues of identity that inform strategies of investment in social position. The ideas developed here resonate closely with those developed in Chapter 2 concerning the relationships between homes – as domestic spaces and domestic cultures – and individual selves. They also connect with the explorations of cooking and cleaning as practices of care and, in particular, with the detailed case studies set out in Chapter 5. Both consumption and care are integral aspects through which many facets of cultural, social and political life may be viewed (Strasser, McGovern and Judt, 1998; Tronto, 1993). In this chapter I discuss the use and exchange value of emotional and technical capitals, expanding Bourdieu's (1984) threefold characterisation of capital – economic, social and cultural – to examine the significance of family practices of consumption and care as assets for changing the material bases of existence. The discussion focuses on two of the ethnographic case studies to illustrate connections between family practices and personal assets as resources for social positioning. The material environment desired and achieved (or not achieved) by the individuals involve practices as a sort of 'script for action' in the terms developed by actor-network theory (ANT) (Akrich, 1992).

The use of 'care' requires some introduction. A distinction made by Joan Tronto (1993) between care as practice and care as disposition frames this analysis. The practice of care refers to the material accomplishment of tasks and activities, while the disposition to care signals the emotional investment in caring. In most cases disposition and practice are bound together, but the distinction between the two concepts allows for a focus on the value of care – and the devaluation of caregivers. The distinction between practice and disposition is, of course, also constitutive of Bourdieu's concept of the habitus. However, in his definition of the habitus, as a structure of disposition, which predisposes individuals to certain practices (choices and actions), there is no room for the emotional aspect of social actions, as I noted in Chapter 2 (cf. Sayer, 2005). Bourdieu is attentive to subjectivity when concerned with the relationships between the individual and the collective, but in his framework disposition refers to schemas of classification, expectations and norms.

Linked to this concept of care, my understanding of emotional capital refers to a capacity to connect, involving acts, intentions and sentiments. It refers to moral thinking about personal connections and intimate life, related to the self and to others, and also being essential for a reflexive self. The concept of emotional capital is, of course, different from that of emotional labour (cf. Hochschild, 1983), which is more commonly found in the literature on care. While a person can use, buy and hire another person's emotional labour, emotional capital cannot be exchanged in similar ways. As with social and cultural capital, it is an asset that derives from personal abilities, connections and investments in and from the self. Like other kinds of Bourdieusian capitals, emotional capital has use value and exchange value in particular markets (see Reay, 2000; Sayer, 2005). I suggest, in line with Lisa Adkins (2002), that reflexive practices, here including the emotional, constitute a means for reclassifying identity, relationships and divisions, thus potentially affecting social position, as the case studies in this chapter will illustrate.

Technical capital is conceptualised as 'the capital of the DIYer' specifically located within skilled workers with technical qualifications. This is presented by Bourdieu (2005) as a predominantly male possession. However, in the context of care, maintenance and display of houses – or of domestic life – more than just the activities predominantly associated with building or male expertise are involved. As I argued in previous chapters, Bourdieu has numerous limitations in his understanding of the contemporary home and the domestic. His theory is more salient for public activities than private ones, which is reflected in the implied gendered nature of his concept of technical capital. Nevertheless, the ability to build or maintain the physical stock of a house defined as a form of capital is not exclusively male. Skills involved in domestic care and maintenance (possessed by both men and women, but exercised predominantly by women), as well as forms of creativity and aesthetic judgements that add value to a property and to the living experience in the home, something not recognised as technical capital by Bourdieu, constitute a form of asset that can be cashed in. Clearly, the décor of the home is significant as a means of 'adding value', which in turn affects the ability to accumulate capital, both economic and cultural (for more on this, see Silva and Wright, 2009).

I selected the Seaman (H8) and Murray-Hall (H18) households for discussion because the forms of appropriation and uses of technologies in their homes illustrate salient issues regarding consumption and care, with different resources of time, money, skills and emotions being employed by them in relation to practices of mundane family intimacy and social positioning. Resources for consumption are generally scarce in both households (see Figure 6.1).

Data reference	SEAMAN	MURRAY-HALL
Sample category	South Yorkshire White Working class	West Yorkshire White and Afro-Caribbean Working class
Family composition	Janet, 35, tutor in cake decoration in community education Daniel, 39, boiler maker in steel factory Megan, 10, state school Sophie, 7, same school, different site Alex, 4, state nursery, childminder	Lynn, 45, unemployed Gillian, 11, state school Halley, 11, state school Chantal, 8, state school Sara, 5, state school
Income (household *net*)	24K (woman = 6K)	7K (benefits)
TECHNOLOGY	____	____
TV sets	3 (L, 8y; 2 Ch's bdr)	2 (L, 9m; Lynn's bdr 10y)
VCR	2 (L, Sophie's bdr)	2 (L, 3y: twins' bdr 5y)
Play station	Yes (Megan's bdr + L)	No
Computer	No	1 (twins' bdr, 6m, 2nd h.)
Modem	No	No
Cooker	G.hob, El.oven 2m	El.hob/oven Creda 4y
Microwave oven	No – had for 8y	Sanyo, 3m
Fridge/freezer	Whirlpool, 2m	Zanussi, brand new
Dishwasher	Whirlpool, 2m	No
Washing machine	Servis, 2m	Hoover, 14m
Tumble dryer	No	No

Key:
L = Lounge
Ch's bdr = Child's bedroom
G.hob = Gas hob
El.hob = Electric hob
El.oven = Electric oven
2nd h. = secondhand
m = month (e.g., 2m = 2 months)
y = year (e.g., 8y = 8 years)

Figure 6.1   Characteristics of the Seaman and Murray-Hall households

The beautiful kitchen[1]

The Seamans' house was in a working class area in a Yorkshire town, in a lane with small, relatively new terraced houses. A couple of pubs, a small church and a corner shop were the local public facilities. I first went there during the school half-term, when Janet Seaman was off work and the children were around, and she had more time and flexibility to accommodate me. I arrived just after lunch. Daniel would come home at about 5.30 or 6.30 pm. Janet (35) had two siblings living nearby and kept in touch with them, but rarely saw her parents, who lived five minutes away. Relationships were strained. She had a further education degree in catering and a part-time job (16 weekly hours plus two evenings) teaching cake-making and decoration in adult education. She was married to Daniel (39), a 'joiner with metal' in a nearby factory. They were both from white, working-class long-time resident local families. They had three daughters: Megan (10), Sophie (7) and Alex (4).

When I arrived, on a cold, wet and grey day, I knocked at the front door but, through the glass panel, Janet shouted for me to come through to the kitchen door at the back of the house. The younger girls opened the door for me, Janet came in and immediately said I needed to excuse them as the kitchen decoration was not yet finished. I looked around and saw myself in a most beautiful room.

The floor was an upmarket imitation of light-coloured wood; the fully integrated cupboards were painted an ocean colour with steel handles and fixtures. The only visible appliance was the hob, cooker and hood, all in steel, fitted with a large steel panel on the wall behind. The ceiling was of a darker colour wood with ten inbuilt directional lights showing just the chrome/steel and flash of light. A large oak farmhouse table was in the centre of the room, with eight matching chairs around it. On the table was a very big glass vase full of gorgeous flowers of delicate colours. The chairs had cushions in a blue-green tartan fabric. The rug under the table was from Habitat[2] with a sand green-blue geometric pattern. A bowl of fruit, a fish tank and a few decorative plants all seemed well placed with matching colours and patterns. A glass window over the steel sink overlooked the washed clothes drying on a line outside. It was true that one small wall, with a blue painted radiator, had not yet been wallpapered. I took an interest in the decoration and praised Janet for her good taste. I stayed in this kitchen for about six hours. Janet made sure the door leading to the lounge remained shut. I was not meant to see the rest of the house. Making me come into the house through the back door had been part of Janet's narrative of herself. This was part of how she wanted to show herself to me. Everything looked sparkling clean.

In most native Yorkshire homes it is usual to receive common visitors and tradespeople by the back door. The front door, which is rarely used, is reserved for important events. I assumed that I was an ordinary visitor, but it felt too intimate to me to appear at the back door. This was also an effect of the design of the house, at the end of a terrace. But there was a contradiction between being taken in naturally and being conspicuously kept out of viewing the adjacent room.

I did see more of the house. I asked to go to the toilet just before I talked to Daniel, at about 6 pm. The lounge was about two-thirds of the size of the kitchen. It was shabby by comparison. The bathroom on the first floor was small. There was just one toilet in the house, which was common in terraced houses of this kind. Although the rest of the house looked clean, it was not as tidy and sparkling as the kitchen. The girls each had their own bedroom, two of which had been very ingeniously created in the loft upstairs, as a result of a large amount of work by Daniel.

Back to the kitchen: why is Janet's kitchen important?

First of all, it was important to Janet. Secondly, the kitchen was important for the relationship between Janet and Daniel and their individuality. It expressed fairness in using the household resources to fulfil personal desires.

Daniel said that he had his 'toy', a Land Rover Discovery bought new the previous year. Janet was entitled to have her toy: the kitchen as she pleased. The kitchen had cost £9,000 in 1999. He had bought it by extending the house mortgage, which was 'not high', about £50 a week (£220 a month).

Although not mentioned by them, it is important to note that the cost of Daniel's toy was much higher than Janet's. However, Janet had the use of the Land Rover during the week for the school runs and to go to work. Daniel drove to work in a Peugeot 205, a car more than 16 years old, which was officially Janet's. Equally, the kitchen was not solely for Janet's use. Yet, Daniel had simply gone along with Janet's project. He did not like the kitchen; he preferred the old one and was still 'struggling' to find where things were.

The new kitchen had been in place for just two months. Daniel's contribution had left big marks. He found the sheet of steel to go behind the cooker. Janet wanted one, it was going to be small, he suggested a place where she could get it from and she got it for £50; it would have cost £400. Daniel also did all the electric installation, following Janet's style instructions. The oak farmhouse table had been a 'present' Daniel was given in exchange for making a steel window frame for a friend's shop. 'It cost me two afternoons in the garage, doing something I like. This is over 300 quid, it is.'

While Daniel, with whom I talked last, was very forthcoming in his accounts and views of the kitchen and its artefacts, Janet's kitchen story was not volunteered. Perhaps she did not expect or wish to disclose the story. It only came out because of my interest in talking about the appliances in the kitchen, which reveal the salience of my research design for the generation of knowledge here conveyed.

We began by talking about the cooker. The kitchen story emerged when we moved to talk about the fridge, which I could not locate because it was disguised by a fully fitted door (and there was another door of the same size!). 'I had to wait for two weeks for them to arrive; the doors were not the right size', she said. 'Oh! I see', I nodded while she opened the fridge door, disclosing that it was a Whirlpool. 'I guess the freezer is Whirlpool as well', I said. 'Yes, it is', she confirmed:

Janet: My old fridge-freezer was Bosch, it was new, I wanted them, they were expensive. But when the kitchen was fitted the Bosch didn't fit. I complained and [the Fitted Kitchen Company] said we needed to wait for three weeks for a new kitchen. We had already had problems with this fitting.

Elizabeth: What kinds of problems?

Janet: We were given a choice of three days and I chose Monday. We were going to have our own fitter. I booked him and I arranged for the old kitchen to be taken down on Saturday, they were coming to do the floor on Sunday so the fitting could happen on Monday. On Friday [the Fitted Kitchen Company] phoned saying they had to postpone delivery for three weeks. We had lots of phone calls and they said they would deliver half of the kitchen on Wednesday. I paid the fitter from Monday to Wednesday for him just to stand about. It was supposed to come on Wednesday at 10 am, it came at 4 pm. We could do nothing. So, the fitter worked very hard … I paid £650 for the fitter for five days, but we were still waiting for units to come. Units were wrong, damaged … I thought if we got a company with reputation, like [the Fitted Kitchen Company], we would be OK. How wrong we were. Apparently it's very common practice. I have now taken them to the Small Claims Court. I need £400 to cover the costs of the fitter.

Elizabeth: What a bad story. And what about the fridge-freezer?

Janet: Well, they gave us these ones. Daniel made them give us. He said he wanted everything taken down. I liked the Bosch one; inside, the shelves were glass and it looked very nice inside. We had more space in the fridge. Now we have more space in the freezer but this is not what we need.

Concerning the other appliances, Janet had got what she chose because she did not buy them from the Fitted Kitchen Company – 'I shopped around'. And she found a number of good bargains, she was pleased to reveal. By then the kitchen story was flowing and Janet was telling me of an achievement and of a story that had taken over her life for a while and was still very important. Her oven cost her £380, the hob £200. At Lewis' they were offered for £650 and £400 respectively. She also paid half price (£400) for the dishwasher, but this was under a special offer from the Fitted Kitchen Company. The washing machine she wanted (Bosch or AEG) would cost £200 more than the Servis she got for £500. As with the dishwasher, she was limited in her choice of washing machine by her desire to have a fully integrated kitchen, which would show no evidence of appliances by hiding the knobs and displays, as well as the body of the machine.

There was neither a tumble dryer nor a microwave oven in the kitchen. However, Janet had had both in the past. She had a tumble dryer before having any children, more than ten years ago, but she didn't like it. 'I like clothes dried in the air outside. It sounds old-fashioned, but that's how I like it' (see Chapter 5). Janet had had a microwave oven for about eight years, but had given it to Daniel's mother when the new kitchen was fitted, saying she did not use it enough. She had thought of fitting it in with the new kitchen design 'but decided not to bother'. Daniel said Janet thought the microwave oven did not match with the new kitchen and so let it go. 'In the same way she decided we would have one, eight years ago. She never consults me.' He'd 'like to have one ... It would be good for when we are in a hurry and need to defrost things'. Interestingly, Janet's decision was based on aesthetics, which was for Daniel an irrelevant issue. He valued the use of the microwave oven and regretted not having it. For Janet, the possibility of incorporating the microwave oven into the new fully integrated kitchen existed, but she decided she no longer wanted to have it.

Why is it significant that technologies that once existed in the home were no longer there? Classic approaches within the sociology of technology, in line with traditional assumptions of the sociology of consumption, have assumed that citizens are passive victims of advertisers. This would lead to conspicuous and excessive consumption because commodities met 'false' needs. As a consequence, the ideology of individualisation grows, and our homes and our lives become increasingly privatised. In this context, consumers, or the users of technologies in the home, appear as non-active or uncreative beings. How much is missing from the grand theoretical schemes that have been proposed to account for the modern intake of technologies in the home? What is the relationship between people and things in the contemporary home?

We see that Janet and Daniel were making active choices. Brushing aside whether they both agreed or not with each particular choice, we see them having a creative influence in the everyday practices of their lives and in the artefacts used in everyday life. Theories need to account for the active role of users/consumers in shaping technological artefacts and their meaning. A cultural identity is being created at the same time (Hollows, 2008). It was clear from the moment of my entrance into the house that I had been involved in a rehearsal of revelation of a newly-created identity. The fundamental process was not just about buying and using goods. The ways in which objects were employed gave a sense of how people were expected to act, and of the kinds of relationships they had with each other. They conveyed a vision of a moral order of the home, which was not circumscribed by the walls of the house.

The practices involved in Janet's and Daniel's provisioning of the home expressed a responsibility to each other and to others. The stress of Janet's story, when she went about making the plans and spending money, was on her ability and skill to find cheap goods and save money. In her narrative she was not spending the resources of the household, but was storing resources for the household (Miller, 1997). There was no sense of personal indulgence in Janet's choices. Even though it was her choice to have the kitchen renewed, the work expressed her concerns for others rather than exclusively for herself.

The kitchen, as it was renewed, was the space for family gatherings – it was where they spent most of their time together, it was the space for family intimacy. In this place they ate and talked to visitors, the children did homework and played; Janet cooked and did the laundry. The ethnographic research materials for this household also included video recordings (see Appendix 1) and in these Janet and Daniel were seen at the kitchen table checking finances and having an evening drink while the children watched TV in the lounge. That this was where Janet chose to place the resources of the household confirms a gender inter-est with the distribution of the costs of caring in intra-family transfers (Folbre, 1994). It is in the kitchen that the bulk of the family practical care work was done: cooking, storing food, washing dishes and clothes. Emotional care is obviously not disconnected from practical care, but it has its specific concerns. This appeared to be highly valued in the context of creation of a space for talking, feeding and enjoyment in an environment with pleasing aesthetics.

Janet appeared to have a large amount of emotional capital available to her, which was evident in how she was clear about her project of identity change for herself and her family and about the concern of her project with the interdependence of the family members. She used her perception of this ideal of caring and living to build an environment that put across the practical and dispositional aspects of her everyday life. The investment made by both Janet and Daniel, relying to a considerable extent on his technical capital, under Janet's lead, is linked to a strategy of upward social mobility. However, in monetary market terms, the capital would not be cashed in in such a way as to include the family in the desired middle-class environment, because the underlying value of the house and the land probably did not immediately justify the investment in market terms. I will return to this point later on.

Waiting and hoping[3]

The Murray-Halls lived in a council house on a fairly poor council estate in another Yorkshire town. The house was semi-detached, built in the 1940s. Neighbouring houses were similar in look and style. All appeared quite drab. I arrived at 1 pm on an early summer weekday. After ringing the front door bell, I heard the door being unlocked. Lynn seemed quite happy to see me, asked me to come in and, as I stood in the hall, she relocked the front door. With either a real or perceived sense of insecurity with the outside world, she was to unlock it again nearer the time of her daughters coming home from school, and locked it once more afterwards.

Lynn Murray was born in 1954 in a white, Protestant, working-class local family. She left school at the age of 13 and had had a number of unskilled jobs (cleaning, factory work and canteen cook). When she was 31 she met Tony Hall, the father of her four daughters. He was a black West Indian sheet-metal worker. After three months she got pregnant. Her father called her a 'nigger-lover and a whore'. She had not seen her parents or two brothers for 13 years. She never got married. They lived together, on and off, for 13 years. But the relationship had ended. Tony had been made redundant due to a bad back. He did not have resources to pay maintenance but he gave the girls whatever he could in money and gifts.

Lynn's first daughters, twins, were 11 years old. She had not done any paid work since they were born. They had lived in their current house for over ten years. Two other daughters were aged eight and five.

The hallway was cramped and small; a room leading off to the right (the twins' bedroom); the stairway straight ahead and the living room to the left. There was a smell of stale smoke in the hall, a sense of slight messiness as well: a few toys and magazines on the floor. Lynn made some comment that she had not tidied up for me. We entered the living room. It was small: a settee behind the door and two comfy chairs either side of it. There was no coffee table in the middle of the room, but one was pushed to the back wall with a phone on it. A television was in the corner. Behind the door was a shining-white new fridge-freezer. Lynn said that there was no space in the kitchen to put it. She was waiting for the council to rebuild her kitchen. The fridge-freezer looked very much out of place, not simply because of its whiteness or apparent incompatibility with its surrounding furnishings, but because it was so new. Apart from the TV, which was also new, the room felt quite old in style. There was a table pushed up to the window behind the left-hand comfy chair, piled high with food and plates. Lynn said they were waiting for the council to mend the kitchen and get some more cupboards. Until this happened, the table would remain full of items and would not be used as a table to eat at or sit around.

There was a definite sense of *waiting* in the household, determined almost exclusively by the work of the council. Lynn had been waiting for years for the kitchen to be improved. Not until then would she be able to move the fridge-freezer out of the living room. Not until then would she be able to use her dining table for its proper use. Not until then would she feel she had got what she deserved from the council. There was a sense of injustice, and powerlessness, in her voice.

The wall to the side of the table was in fact a cupboard: cream wooden doors, reaching the height of the ceiling. Inside it were food and tins, contents normally associated with the kitchen. Behind the settee and next to the fridge-freezer was a dresser with sliding glass doors, filled with pictures of the girls. Mounted, hanging school photos of the girls adorned the opposite wall as well. There was a real sense of pride in these photos. Nothing else was on the walls. There was a gas fire below these hanging photos. The room felt messy: girls' hair bobbles lying on the floor, magazines on the sofa. The television was on when I entered the room and it was not until we started talking 'properly' (with the tape recorder) that she turned the sound off (although not the TV).

I made reference to the enormity of the television. 'Well, it's costing me a fortune. I got it on credit and the interest is enormous.' It cost 'only £300 and I paid another £70 for the 5-year warranty'. This seemed incongruous with the lack of money in the household:

> They discriminate us people, they do: I can't afford it. I mean, I wanted to hire a digger for the garden. I could have hired it, but they wanted a driving licence, which I don't have, a passport, which I don't have, and I had to have transport to bring it here, which I don't have. They discriminate against people on income support. So I couldn't hire one.

This explained why the grass outdoors was so overgrown. The implication of what she said was, it seemed, that because of the lack of affordability, she was obliged to make an enormous expenditure with the television. This sounded irrational but it made sense in the context of Lynn's life. I remarked how bad that situation was and she replied: 'Well, I hope you never go there.' She sounded as if she were in a 'place' that was undesirable and difficult.

There was a video recorder under the television. She did not know how old it was because she bought it off a friend. Lynn had had it for three years. She paid £40 for it. Her friend taped a lot of things for her. 'To be quite honest, I wouldn't even know how to tape.' They did however watch films on it.

The computer, in the twins' bedroom, was acquired in a similar fashion. 'It's only because I know someone who deals with computers; he's a computer whizz ... I bought it off him.' 'It's not brand new; I only paid £100 for it.' The girls used it. When they first got it, they used it for all sorts of things, but afterwards they just played games. Lynn showed me the CD-ROM drive on the floor; she did not understand what it was for: 'We don't know how to use it!' There were some games already on the computer that they played. Lynn did not understand the computer at all, and she was not familiar with it. She thought the kids were quite bored with it, 'but we're stuck with it now'.

She remarked how quickly her money went on water rates, gas rates, electricity, the phone, the TV and then she would be lucky if she would have £90 left to last her the fortnight. Her fixed expenses were around £220, cigarettes included. Her daughters were expensive, especially the elder two: they wanted all the brand names (Nike, Adidas and the like). If the girls wanted their hair straightened, it was expensive: 'Unless the dad pays for it, I can't.' 'It's a hard life.' The father of her daughters sometimes asked Lynn what she was feeding them. She believed he thought she was not feeding them correctly. The thought of being seen as a non-nurturing mother seemed to hurt her.

Lynn talked about her routines and when the children had baths. Asked if she had a shower, she pulled a face in wishful thinking: 'I would really like one, but a shower's a luxury, isn't it? It's like having a car and a phone and a TV; it's a luxury when you're on income support.'

She then made reference to the daughters' father: 'I mean, the dad's bought the fridge-freezer: I would never have had (a new one).' She took me to it, opening it. The four freezer drawers were very full of budget-type ready meals, but the fridge was fairly empty. There was no fresh or perishable produce in the house. The fridge-freezer had arrived a few days before. It cost £350. I asked why Tony had decided to buy it: 'I think he felt sorry for me.' Also, there was some hope that the kitchen would have been ready by then, so that the new one could be put in. Lynn had a long history of secondhand, quite unreliable fridges. There seemed to be a long trail of hand-me-downs, all of which seemed to have broken and stopped working fairly soon after she acquired them. The story was the same with washing machines.

I asked if we could go into the kitchen. 'Oh, I'd rather you have a look at it when it was all ready and finished. It's awful ... it's awful ... it's a disgrace.' But she let me in. The kitchen was tiny and undecorated. On one side of the kitchen was the sink and draining area. There was very little work surface space (approximately one foot by half a foot). Two kitchen cupboards were mounted on the left-hand side of the wall. There was no other storage space in the kitchen. All surfaces were loaded with items: plates, pots, jars of cereal. There were no dirty plates or cups: things were clean but difficult to see because of the cramped nature of the room. Through the kitchen there was a small space: the once-used larder designed with the house. Lynn was waiting to have this wall knocked down to expand the kitchen so that she would then be able to fit in all of the items currently in the living room.

Cooking seemed unimportant. Lynn had had her 'new' microwave oven for three months: she got a loan 'off the Social' to buy it. It was in lemon yellow (and matched the toaster and kettle in colour). She had had another one before, for about ten years. She used it every day, either for warming up foods or to cook from frozen. She used this much more than the oven: this was her actual oven. The girls knew how to use it: she demonstrated how easy it was to use the non-digital style. She really did not seem to eat very much at all, as I noted in Chapter 4. She appeared to survive on coffee and cigarettes: she would drink about ten cups of coffee a day with milk and sugar: 'It keeps my adrenaline going.' She smoked 20 cigarettes a day: 'I don't do anything, I don't go anywhere; it's my little luxury', she said. But then she volunteered that smoking was a complete waste of money. She enjoyed it, but when she considered it cost her between £25 and £30 per week (in 2000 this was £3.65 for 20: she was very clear on the price), she concluded that it was a bad habit.

Although Lynn hardly ate anything, there was a clear narrative surrounding the kitchen, the council and her interaction with both, as well as the perception of her kitchen to others. She told me a story about how her neighbour, who had never been in the house before ('but I had to see her about something'), thought that Lynn's kitchen was disgusting, not in terms of hygiene, but in terms of how the council was letting her live in it. She also referred to a housing department officer saying: 'If you left here, I'd condemn this kitchen.' The reasons for him saying this apparently were that the cooker was in a dangerous place and the plugs were obsolete:

> He said these kitchens were built not for microwaves, fridge-freezers. Like 40 or 50 years ago when these houses were built, they had the old washtubs and they had the larder cupboard and they washed everything by hand ... they didn't have microwaves, fridge-freezers, split level cookers and all gear that they have now.

In the future, when the kitchen was repaired and redecorated, the fridge-freezer would come in and be put where the oven is now. The oven would move into the old larder. Lynn seemed very clued up as to how she would like her new kitchen. She planned to decorate the whole kitchen in a lime green colour. After the council did the repair job, she would then need to find the money to decorate it herself. Upstairs she had got everything else for the new kitchen: the blinds, a new drainer, a new clock, a knife set, 'everything'.

Lynn's everyday life lacked nourishment, both material and emotional. This was concrete in the kinds and quantities of food available and consumed. It also showed in the stale air, the stale state of most of the artefacts in her home and in some of the structural features of the house. A number of things were secondhand, acquired via networks of friends or from the council. She was bitter about her poor state, she felt neglected, powerless. She made a number of apparently irrational, incongruous expenditures, like her cigarettes, her large screen television and the unused computer. For Lynn, these expenses, which she felt obliged to make, were controversially an effect of her inability to afford things. She was in a kind of 'Catch 22'. As she remarked on her television purchase and her attempt at hiring a digger for the garden, her exclusion concerned various forms of discrimination and the only way she found to get included was by performing as a consumer who had choices. Her 'choice' was to buy a very expensive TV. Aside from her consumption of cigarettes, the incongruity of owning other products she possessed lay in her lack of expertise in using them (cultural capital in its technical capital form). Behind such kinds of 'irrationalities' there was an expression of a desire to compensate for a lack. Lynn had plans to beautify her kitchen: she acquired many new things in the hope of redesigning her nourishing centre. Until then her kitchen was her room to smoke in. In her story she seemed to be waiting forever.

Resources for consuming and caring

Three themes are significant in these stories:

1. Consumption practices, as a critical part of the creation and maintenance of a valued sense of the self, and considerations of use value and exchange value are important in decisions to consume.
2. The mobilisation of resources by individuals as part of their inclusion in particular categories of social class, or simply as consumers, relates to economic, social and cultural resources (or capitals) involving practical skills (technical capital), and to a disposition for personal connection and intimacy related to the self and to others (emotional capital).
3. Differences of wealth, race/ethnicity, employment and marital status, number of children and other divisions need to be explored in terms of access to positions and dispositions revealed in particular consumption practices.

I look at the cases of the two households in relation to these concerns.

Consumption, the self and value

The two homes illustrate that material objects are typically used to construct and present a certain persona and support group membership in the ways in which objects are connected to themselves, the family and consumption practices (Bourdieu, 1984; Featherstone, 1991). Because they are not well-off, considerations of use value and exchange value are salient in the narratives of appropriation of material goods (Warde, 1994). All stories of possession carry a cost tag and a personal acquisition strategy that are intricately linked.

The Seaman story unravels strategies of achieving access and relating to groups which possess those objects required for their project, chiefly friendship and work networks. Getting hold of the objects made them feel good and valued. They took pleasure in showing their resources – technical, social and emotional – and talking about how resourceful they were in obtaining them. There was a certain homogeneity and harmony in the narrative accounts and object displays.

Lynn Murray's home mixed some 'luxury' objects – the big new television and new fridge-freezer – with secondhand, repaired appliances, handed-down objects and brand-new objects saved for a future project of display in her kitchen. This was the least resourced household in my study. Everything was scarce: money, food, emotional nourishment and skills. Lynn had low self-esteem and her horizon of change stopped in her kitchen, which was waiting and waiting to be renewed by the council.

Although the practices in the two homes show links between consumption, value and a project for the self, this is most evident in Janet's narrative. Why did Janet Seaman decide to place household resources into making a beautiful kitchen? Janet's kitchen was middle class and signalled a direction for her (and her family's) social position. I say that Janet decided to make, not simply to have, a kitchen. She did not buy a kitchen, she actively involved herself and used various personal resources, those of her husband, social and business connections, to make the kitchen. Exchange value aspects were very important in the face of her scarce financial resources. All items were purchased for a price below the market value. She built a nourishing kitchen and she felt nourished with her project. There was an enormous use value for the family in the project of the kitchen, both because of the ambiance of a nourishing family life, and because of the supposedly middle-class ticket achieved with it. She was able to present herself with a new desired personal identity, via her kitchen. Yet, this kitchen differed greatly from the rest of the house. It signalled a direction, a motivation and a tense fragmentation of the self. The senses of belonging appear in tension in two ways.

Firstly, Janet belonged to two different places: her working-class origins, to which her house location and the decoration of the other rooms testified, and the middle-class location she wanted her family to achieve. It seems appropriate to recall some criticisms of the ideas of the power of a reflexive self in the aesthetisation of everyday life (Giddens, 1991; Featherstone, 1991). In relation to how people deal with constraints in his study of kitchen renovation, Dale Southerton (2001) notes that reflexivity of social identity in terms of class position was more relevant than reflexivity of self-identity. Reflecting about the limits of the individual's power of reflexivity, Derek Layder (1996) argues that it is the individual's power of motivation that turns them into architects of action (this echoes Thévenot's (2001) notion of creativity in the constitution of practices). While Lynn had little of this power of motivation, Janet's reflexive self designed the kitchen project in relation to a social identity which motivated her to drive it into a concrete enterprise. But the overall enterprise was limited and limiting. Janet could not go further in her relocation project. While the use value of the capital she invested in her kitchen was high, the exchange value would not allow her to invest in a middle-class location for a project of a new home. The range of the family's cultural capital – technical and emotional – was too limited for a projected change of social position.

Secondly, Janet's narrative of self and family was fragmented in a tense manner. My confinement in the kitchen meant that Janet did not want me to see the other rooms of the house. The contrast between the hidden rooms and the room on display was more significant because of her efforts to keep the door shut. In addition, she did not volunteer to tell me her story about the kitchen. This turned out to be a story of struggle to pursue and to achieve her beautiful kitchen. It revealed a very resourceful Janet, but this was not the Janet that she wanted to have revealed. Both these two hidden things showed the short reach of her project, that her desired social inclusion as middle class was very partial. Janet knew it, but she did not want me to know it, or she did not want to know that I knew it.

Inclusion, emotional and technical capitals

The idea that consumption is a site for the creation and maintenance of a valued sense of the self, but that narratives and performances of selves are also linked to differential access to goods, links the politics of consumption to the availability of economic, social and cultural capital. Yet, these concepts of capital need to be refined to account for the development of capacities to enhance resources from the application of skills and to choose and achieve particular ideals of connection and intimacy. The ability to apply one's practical skills to making something, to acquire a new object or to a new situation, and a disposition to engage emotionally with the self and with others are all assets that, like capital, relate to social positioning. These are linked to specific strategies of advancement in social space.

Concerning emotional capital, in contemporary narratives of the self, the responsibility for maintaining relationships is increasingly less restricted to women. Men are expected to care. The overall moral sense of responsibility in provisioning the home was not gender-specific in the narratives of the families in my study, but the practices expressing particular senses of social inclusion or exclusion seemed particularly relevant for the women.

In the Seaman household, the kitchen renovation was Janet's project. Daniel simply went along with it, although he was still a bit unhappy with the change. I remarked that Janet's kitchen was not just about family life. She clearly wanted a space for intimate family living but her project was also about social place, showing where she wanted to be included, what she was claiming to be a part of, together with her family. She had a strategy of investment which carried her family into a more valued social space. The story shows that both Janet and Daniel followed particular capital accumulating strategies on the basis of their available technical and emotional capital.

Race was a feature of social exclusion for the Murray-Hall household. It negatively affected the family's emotional capital. Because of Lynn's link with a black man, her own family withdrew any material and emotional support. Race was a highly relevant issue in Lynn's narrative. Its tension was revealed both in how it was actively engaged with and in its non-explicit role. Race appeared in Lynn expressing lack of support by being discriminated against by her father – and implicitly her mother, account-ing, by inference, partially at least, for her poor emotional capital. Yet, race was an issue that did not appear in Lynn's home. It exerted the power of an 'un-talk-about-able' (Morgan, 2004). She had never talked with her daughters about her own parents, about why the grandparents had never seen them. Interestingly, although the girls had visited their West Indian grandmother, Lynn had had no contact with her. It was as if Lynn were up in the air, not belonging to either white or black groups, disconnected from all.

Lynn's exclusion from consumption was experienced as an undesir-able and difficult place. She felt incapable of having or of being someone of her choice. This affected her sense of self-worth, which led to a sense of greater exclusion. In reality Lynn belonged to a circle of friends, allow-ing her some social capital and giving her access to consumption. This access was to second class, handed-down, cheap goods, and she did not appreciate them. However, she appreciated her apparently 'irrational' consumption of (for her) luxury goods: the new big television, the new fridge-freezer and her cigarettes.

Differential assets and particular consumption

This analysis of consumption practices in the home is concerned with the interiority of the family relationship. It shows individuals (most particularly women) operating as conscious agents of change in the sphere of intimacy and also in relation to social divisions. Both families are cases of relative economic scarcity, but there is a considerable difference of wealth between them. They present different ethnic/race affiliations, which bear on their senses and practices of position in social space.

Lynn's case stands out as being the most deprived of assets for consumption. Hers is a case of a white lone mother, with mixed-parentage (Afro-Caribbean) children, living on income support and having a larger number of children. But a comparison of the list of technologies in both households does not discriminate against the Murray-Halls compared to the Seamans. Actually, they have the newest television and the biggest, and newest, fridge-freezer. This raises the moral debates about choices of purchase. In Lynn's sense of morality it was because she could not afford to hire a television that she needed to make an enormous expenditure to buy one. This is simultaneously an act of revenge and of victimisation. Perhaps her cigarette consumption would fall within a similar category.

It seems important to recall some of the arguments from Chapter 1 about how homes are differentiated units of consumption and to emphasise that significant divisions occur within the home (Jackson and Moore, 1995). For example, whereas Lynn resourced the everyday, the father of her daughters paid for 'luxuries' like straightening the girls' hair or supplying a 'new' fridge-freezer. Also, the cost of Janet's kitchen was much smaller than Daniel's car, as I noted earlier. Gender imbalances are significant in consumption practices (Casey and Martens, 2007). The balance between differential possession of assets in the home and individual consumption may however be levelled out if a sense of fairness prevails. The next chapter focuses on these ethical dilemmas of the give and take among family members, reflecting on the interplay between individuals and objects in practices of relating in everyday life.

7
Domestic Dilemmas

The basic assumption concerning the constitution of social space is that individuals are located relative to others. Accordingly, the consideration of other people and the positioning of the self and of others – or of 'things' – involves classification and judgements. For Bourdieu, social position is predicated on the social hierarchy of class which conditions all matters of engagement in social life. Expanding on Bourdieu's framework, I note that aspects of identity based on gender, sexuality, age and ethnicity are also relevant for social position. For Latour, connections are not predictive and these are traced as networks emerging from actual practices and configurations. As noted in Chapter 1, I both contest Latour's loose and flat conception of the social and the determining aspects of Bourdieu's ontology. In the looser conception of social ontology that I embrace, the world appears as fluid and dynamic. At the same time that tracing connections is an empirical matter, connections have causes and effects and constitute – as well as being constituted in – patterns. An integral element of relationality concerns moral issues, since justifications for how things and people go together are integral to human practices (Sayer, 2005; Thévenot, 2001).

I explore these ideas, engaging in this chapter with the 'moral choices' made by women and men in matters of intimate everyday living in the home. What issues are involved in choices about who cares for what, whom and when? What can one ask others for? What should one negotiate about? Why? How do objective conditions relate to subjective choices? What impositions do the material resources of everyday life place on domestic dilemmas?

My exploration engages with sociological analyses like that of Zygmunt Bauman (1995), and his concern about the increasing need of individuals to become more moral in post-traditional societies because set rules have diminished, and also those of Luc Boltanski and Laurent Thévenot (2006), who claim that 'justifications' of one's actions to others are an essential dimension of social living. But my understanding of morality, in line with the feminist literature on gendered ethics, points more firmly towards practices and concrete actions. Following Joan Tronto (1993), whose framework was central for my analyses of practices of care in Chapter 6, the morality I refer to is based on moral sentiments, not on reason. This means that morality is not formal and abstract, but that it is tied to concrete and material circumstances, rather than being based on a set of principles; it is an activity grounded in the daily experiences and moral problems of real people in their everyday lives (cf. Gilligan, 1982). I consider 'social' morality (Finch, 1989) as a wider frame of reference for the ways in which people live their lives under particular social circumstances where gender, class, race or ethnicity are key categories of moral thinking. I am also mindful of the psychosocial dynamics related to capacities to care explored by Wendy Hollway (2006), particularly regarding the projections of and identifications with the dilemmas of others. With this exploration I seek to understand some of the social patterns implicated in these everyday moralities.

Selma Sevenhuijsen (1998) has stressed that in practically all contexts of behaviour, individuals are confronted with questions of judgement and responsibility. But if reflection on 'how to act' is related to personal agency and to interpretation, this does not happen in a vacuum. Clearly, the work with morally thinking and acting persons requires a conceptual framework of moral subjectivity which is marked by positions in social space (Sayer, 2005). Janet Finch (1989) (and with Jennifer Mason (1993)) has stressed that moral values are not simply political or cultural conflicts, but that ideas about moral obligations derive from the wider culture *and* from personal circumstances. People mobilise obligations through the responses of other people. An individual's understanding of the social order helps to reproduce it accordingly because, in acting, she or he makes specific claims about which rights and obligations are going to be recognised in a particular situation. Normative guidelines about how one should act have a role in the process of negotiations, which is not necessarily free from coercion, persuasion and manipulation.

This patterning of social hierarchies in moral reasoning is further stressed by Michèle Lamont (2000), who found that people with relatively low cultural capital often emphasise moral ideals, as boundaries of identity, around issues such as the importance of personal integrity, maternity, family and friends, because these give them a sense of dignity, or even a sense of being *better*, than middle-class individuals who, from their perspective, often lack or compromise in these areas.

Searching patterns in everyday moral practices

Katie threw some salad or vegetables into the ready-made meals so that she would feel less guilty about feeding them to her family. Rena ironed every garment to demonstrate her care for the clothes in her home, except that she did not launder for her 17-year-old son after he complained about the quality of the washing. Colin found that getting involved himself with cooking or cleaning was entirely out of order, even though he cooked a Sunday breakfast which his children really liked. Gabriel arrived home after his children were in bed explicitly so as not to be disturbed by their boisterousness or demands. Diane spent as little time as she could on domestic tasks, purposefully mindful of the fact that she worked long hours away from home. And so on ... individuals take decisions according to personal agendas and to what their current circumstances and partnerships entail. Examination of the living patterns of earning, consuming, allocating time and caring of the families in my study shows that these were particularly defined in relation to their circumstances in the labour market and the interaction between partners. The patterns also involved reflexive choices about how to deal with issues that related to individual life circumstances, personal and social resources, and contexts. They represented 'moral choices' (Silva, 2004).

The purposeful investigation of morality in my research was based on the use of a set of vignettes, although moral concerns were also evident in most of the personal and relational matters discussed in this book, such as the use of space in the home or uses of time, practices of cooking and cleaning, and care practices and dispositions. This method involved reading stories in the form of vignettes to research partici-pants and asking their opinion about events. I started with my own set stories, but this triggered the telling of further related stories involving self, friends and family. Here I focus on the moral orientations of the adults in relation to gendered divisions of labour and use of resources in the home, with reference to three scenes. These are long vignettes about hypothetical characters – a woman and a man with children – in a particular circumstance. The interviewee, upon hearing the story, was invited to respond to the characters' dilemmas. As a method, the use of vignettes recognises that meanings are social, while allowing an interviewee to define the meaning of the situation for herself or himself. In addition, it assumes that morality may be specific to particular situations, bringing out both the personal and the public morality of relationships. In my vignettes I used a method of building in a temporal dimension and altering the circumstances of the hypothetical characters at a later stage. Individuals were invited to make a choice about what 'ought to' happen in the first and second stages. What happened in the second stage was independent of the interviewee's view of the first stage. This increased the complexity of my stories.[1] The vignettes were always read out at the end of a long conversation about the home, activities, technologies and personal life.

In the three situations discussed in this section, negotiations often appear implicated in differences of power based on hierarchies of gender and class positions and presuppositions about them by individuals placed in society and culture. The scenes refer to dilemmas of three women – Jane, Emma and Susan – in relation to what to do for themselves and others in their everyday home life. The stories focus on a conflict situation. For Jane the conflict referred to recurrent episodes, for Emma it arose from a change of a way of life, and for Susan it was a repetitive daily affair. The three women were each mothers of two children. The ages of Emma's and Susan's children were known and it was made clear that Susan had a heavier burden than the other two. The cases of Jane and Emma raised some aspects of technology use in the home and those of Emma and Susan suggested possibilities about the gendered organisation of housework. The three cases raise issues about what to give to, take from, ask for and expect from a partner. I constructed some similarities between the stories but I also made them progressively more conflictive. On the whole, interviewees very quickly identified with aspects of all of the stories. Some issues brought up complete agreement about what should be done, while in others, opinions were clearly split along gender lines and it was often possible to discern how personal concrete life circumstances had a strong bearing upon the opinion offered. This is why identification was also so strong and prompt, revealing in particular gender and class issues.[2]

Recurring compromises

Jane and Andrew had been married for 15 years and had two children. Jane wanted to buy a VCR. Andrew did not. Eventually they acquired one. At first Andrew ignored it. But Jane learned how to operate it and used it very often.

One day Andrew wanted to watch a football match but was unable to do so at the time it was shown. He asked Jane to record it for him. Jane did it. Then it happened a second time and Jane thought Andrew should learn how to record programmes and taught him how to do it. But instead of doing it himself, Andrew kept asking Jane to record things for him. One day she refused and Andrew claimed that, since she wanted the VCR, she ought to record programmes for him. Do you think that:

(a) Jane should maintain her refusal?
(b) Andrew was right and Jane should record the programmes?
(c) Jane should carry on doing it until Andrew learns to operate the VCR?

I often probed responses by asking 'why?' following the choice made. Before talking about the vignettes, I had invariably talked with each person about the technologies in the home, and the use of video recorders, computers and the Internet had been particularly addressed. The issues in the vignettes had often already emerged in other ways. Comparing statements of partners, there were collusions, false accounts and contradictions. Of course, the vignettes only disclosed beliefs, but they triggered accounts of actual happenings or complemented stories previously told. The fact that the vignette refers to a technology owned in nearly 90 per cent of households at the turn of the century but is decreasing in importance should not affect the significance of the dynamics portrayed. Objects mediate relationships creating or solving dilemmas in different ways.[3]

This vignette deals with gender, the use of resources and claims on a partner's expertise, time and interest. The elements of the story include: (1) a couple in a long-term relationship with children; (2) a conflict over purchase, followed by agreement and acquisition; (3) female technical expertise, male non-expertise and male demands for female technical services; (4) female refusal resulting in conflict. The story at this point ends ambiguously. What views emerged? Most men (over three quarters) thought Jane should have carried on recording programmes for Andrew until he had learned, but most women (nearly half) thought that Jane should have refused to carry on doing it. No one thought Andrew was right in thinking that Jane should have carried on tape recording for him indefinitely. But opinions were not gendered in traditional ways:

It's incredibly annoying: this was the situation I was in with Karen [a former partner]: she decided she couldn't do it, so I had to do it. ... she couldn't be bothered to learn ... it makes you nervous because you think if it doesn't work, I'm going to be in trouble. It's not just men who do this! (Rebecca, H17)

It's like he's saying to her: 'it's yours'. Like: 'You've had the kids, you should do all the chores for the children ye know, I'm just sort of in the background and I don't want to learn to change nappies ...'. You have to share the jobs. In a situation with me I'd say to Phil ... 'I feel you should do it'. (Chris, H13)

... having a situation similar to this in our house where Robert doesn't know how to programme the recorder and I'm the one who does it, I don't mind doing it ... but if ever he complained to me ... I should say: 'Well, that's it, you learn to do it yourself!' (Frances, H11)

Issues of masculinity emerged in that while some mentioned that there was role reversal in the story, others identified with Andrew, remarking on a possible technophobia, in some cases undisclosed. Robert Gibson (H11), for instance, did not tell me that he did not know how to record videotapes. I asked him why he did not tape things. 'Cos I'm not normally here for what's recording', he said. I suggested 'But you could then pre-programme the video', and he replied: 'Oh, Frances does it.' Yet, he thought Andrew should have learned.

I ended up doing stupid things, like Andrew, because I wanted to cover the fact that I didn't know how to use it. (Richard, H20)

Some found it uncomfortable to disclose their lack of knowledge relating to technologies. Technical 'incompetence' was not just a problem for men like Robert above. At the Mitchell household (H7), Nancy did not tell me why her husband had bought a new VCR with 'video plus' ('so easy, you just tap in the number of the programme from *TV Times* and it's done for you'). According to Alfred, however, 'She wouldn't learn how to do it in the old one' and this is why he bought the new VCR with 'video plus'. But Nancy's comment on the vignette was, in the light of her husband's comment, one of faked technological ability:

Well, it wouldn't happen in our household, whatever we buy we've all got to know how it works really ye know, especially – but they're so easy to use and – we've had videos for years so – I couldn't comment on that because it wouldn't arise ...

Mostly, partners had different views about what should be done, remarking that everyday morality in households is commonly diverse, with individuals taking different views about what to give, what to take and what to ask for in particular circumstances. I will come back to this point later.

The second part of the vignette presented a resolution of the previous dilemma and brought in a time dimension and a newer dilemma.

Eventually Andrew learned to operate the VCR.

One year later Andrew decided to buy a computer with an Internet connection. Jane was not too keen on the idea but accepted it. After the computer was first installed, Andrew used it all the time. But then Jane learned how to use it and discovered many things that interested her on the net. In the evenings she would spend time navigating instead of watching TV with Andrew or keeping him company while he was having his dinner. Andrew resented this:

(a) *Do you think that Jane should not use the Internet and should keep Andrew company?*
(b) *Do you think that Andrew and Jane should come to an agreement about when they each should use the computer?*
(c) *Do you think that Andrew should veto Jane's use of the Internet?*

Here, the elements of the story are: (1) a new potential conflict over a purchase, followed by agreement and acquisition; (2) male and female technical expertise; (3) female concentrated focus of interest away from her husband; (4) the husband's resentment. What views emerged?

People were overwhelmingly in favour of Jane and Andrew reaching an agreement. Only Brenda Addison (H3) thought that Jane should not use the Internet and should keep Andrew company. Two men thought Andrew should veto Jane's use of the net. Interestingly, they were both self-defined 'technophiles'. One, Scott Bird (H9), was married to a woman devoted to supporting his needs. The other, Ronald Chambers (H5), who had a relationship committed to gender equality, identified with Jane in gender role reversal: 'I identify. Sometimes I will be so tired she would watch TV, I leave her and go surf the net – it takes me away – ... It's a very real problem. – But, what do I think about this? ... the simple thing to do is to negotiate time.' His wife Rose also identified with the situation: 'We had the same problem. We talked, now I say when Ronald spends too long with the computer.'

Mike Goodman (H6), living in a relationship of traditional gender divisions, said he 'wouldn't restrict, apart from the cost ...'.

Fear of addiction was also a reason why it was thought that use should be curbed even if in a negotiated, not imposed, manner. Irene Hays-Field (H14) said:

> ... [the Internet] it's so addictive ... you hear about internet widows and widowers and you can see why that happens so ... go on it for a certain length of time and then make yourself come off it 'cos I think it is an addiction in many ways, just like television can be.

Another negative outcome, expressed by Irene's husband, Ian, might be divorce. Ian took a long pause, grumbling, smiling and then said 'I think Jane should use it less':

> *Elizabeth*: Should Andrew ask her to do that?
> *Ian*: No, Andrew should realise ... he would go out to the pub, leave her to it.
> *Elizabeth*: He should go out to the pub and leave her doing the net?
> *Ian*: Well, yeah, the problem seems to be that he wants her company and she's not letting him have it, but his company is available to her if she wants it so ... remove that availability.
> *Elizabeth*: And you think that ...
> *Ian*: This marriage is going to end in tragedy!

A similar view was presented by Daniel Seaman (H8): 'They don't seem to have that much in common, do they? I think there's a divorce in there!'

This idea of divorce and separation resulting from a domestic conflict will recur in the other two cases. The risk of non-reconciliation is very present in the minds of some people. Domestic problems, as unresolved dilemmas, reflect the seriousness of the ethics of the everyday. Responses also stress the need to change and to achieve a comfortable situation between people living together.

Changing a way of living

The vignette of Emma and John relates to the change of the woman's employment from part-time to full-time and its implications for domestic arrangements. The first part of the vignette reads as follows.

Emma and John had two children who were seven and nine years old. Since having the children, Emma had worked part-time. John worked full-time. They shared housework and childcare in a way that they felt was 'fair'. But now

Emma decided to work full-time and she found a job. She felt that she and John needed to reorganise their shares of household work. She was not clear as to what to do.

I will tell you her options and I would like you to say what you think she should do:

(a) *Keep things with John as they are by not demanding more from him.*
   (a-1) *She can save time on her work by buying appliances such as a dishwasher and a microwave oven.*
   (a-2) *She will then simply leave undone things that she cannot do herself.*
(b) *Change the way things are with John by asking him to take on some new tasks and check what can be left undone.*

The dilemma emerges out of Emma's choice of working full-time. Should she absorb all the costs of her choice on her own by 'keeping things with John as they are' or change things by implementing a new division of domestic labour? The majority would 'change the ways things are'. However, over one-third – equally among women and men – would 'keep things the same for John'.

'If she goes back to work it's at her own peril. She's the wife. She had the children. It's really the men who get jobs', said Rosanne Goodman (H6), the only full-time housewife in the sample. Her husband, Mike, similarly reasoned: 'If I thought my wife was going out to work and … at a price … there was some inconvenience … then I'd say yeah, "buy that piece of technology and you can continue working".'

Rena Rock (H10) said: 'She should not put any more pressure on him. Or he'll be doing nearly all. He's the main breadwinner.' John, in affinity to his wife, reminisced about his experience of her taking a job two mornings a week over the summer:

> We had a similar situation ourselves … I said: 'You can go to work if you can cope with everything else.' I don't want to come home and find there's no tea or my shirts aren't ironed. But … we found that … well the place didn't get hoovered, things didn't get done.

Trevor Lakin (H4), who had been unemployed for five years, married to a woman who worked as a cleaner during school hours, told me 'Well, I certainly don't think she should put any more strain on her husband':

> *Elizabeth:* Why shouldn't she?
> *Trevor:* Why shouldn't she? He is working isn't he? He's working full time and she was working part time, then full time … coping

with young children. That's where a childminder comes in ... full time. See, I'm probably an old school, I'm one of these people that think I'm the breadwinner ... Emma wants to go ... Emma wants to go to work full time she's gonna have to make some arrangements for the children. I don't think that should fall on me.

Those who voiced conservative gendered home arrangements appear not to give Emma a right to choose to have a full-time job. These opinions echo a puzzling question as to 'why is she going to work?', as Uli Naylor (H15) asked. However, I stress, these are not the views of the majority, who thought that Emma and John should change arrangements to share more.

'It would be unfair for me to ask Diane to do it now when she works more. It can feel liberating [for me] to brillo pad a pan after a long day's work', said Marc Churchill (H16), the senior academic, who worked from home. Similarly, for Richard Bartholomew (H20), lone father and a PhD student, working from home: 'Once you start working full time, you need space at home as well ... work isn't just what you do when you're there ... and it's fair that that gets sorted out because in that position he already has that space: it's only fair she gets it too.'

The second part of the vignette unrolled independently of the views expressed.

Emma proposed that John should take a greater share of housework and child-care and he agreed. But after about six months Emma felt she was doing as much as before and her share of the work was not 'fair'. What do you think Emma should do:

(a) Talk things over with John again and get him to do more?
(b) Get some paid domestic help?
    (IF (a-1) HAS NOT BEEN SELECTED):
(c) Buy a dishwasher and a microwave oven, and perhaps other appliances to help with the housework?

When the situation moves on, six months later, nearly half of the interviewees, but a higher proportion of women (about three-quarters), favoured a multiple strategy. Women would primarily seek paid domestic help as a solution, but half of the women would also buy newer domestic appliances to help with the housework. On the whole, women did not think that by talking to John again, Emma would get her problems resolved, although in general they would keep talking. However, most men would like Emma to talk to John again, followed by a nearly equal proportion of men who also favoured the hiring of some domestic help:

> *Frances* (H11): ... it may be easier to get some paid domestic help.
> *Elizabeth*: Why is that?
> *Frances*: Men have the best intentions. They say they'll do this, that and the other but soon go off the boil ... This is how we got the dishwasher. It was Robert's job to wash up ...

> You always think you do more than the other person. If he thinks he's doing more he's not going to do any more. I'd still talk things over with him. I'd just get help for the sake of domestic harmony ... I'd not stop talking about it. I'd get a cleaner, but also talk. (Rose, H5)

> Talk again and again. You can have all the technology ... without the talking there's no solution. (Marc, H16)

Individuals' identification with the situation was striking. Very quickly and easily, they referred to what they would do and personal examples were offered. Conservative and also more contentious views also emerged. As before, Mike Goodman (H6) makes a good illustration: 'This [paid domestic help] is an option to be seriously considered, bearing in mind the cost doesn't exceed the other income.' As I noted earlier, Mike had also mentioned 'cost' in the context of the first vignette as an issue concerning whether Andrew should restrict Jane's use of the net. His opinion about Emma's choice was in tune with his wife's: 'It may not be fair for the husband to have chores ... I was thinking of my husband saying: "Is it worth it? By the time you pay for housework ... there's not much left".'

Scott Bird (H9) said that 'Domestic help is the easy option' or that she (Emma) should 'leave work'. Yet, he also said: 'I don't like the idea of someone strange wandering around my house.' Luckily for Scott, his wife Wendy was happy to look after the children and home, working as a tutor from home despite having got her qualifications as a primary teacher at the same time as he did his, 17 years previously. Scott was a successful computer system designer at the time of this conversation. In social space their respective occupation status differed widely, though they complemented one another at home.

The cost of paid domestic help may be seen as curbing the prospect of a woman's employment, but it may also be regarded as a means to 'punish' a male partner for the woman's 'need' to go out to work, as expressed by Lynn Murray-Hall (H18), the unemployed lone mother: 'If [I were working] ... I'd say to him: "You've got the money, so you pay for a cleaner to come because I'm not doing it all..."' Lynn implied that Emma was obliged to take on the job. In her fantasy, the male authority prevailed when she angrily delegated to 'him' the power to hire a cleaner. Also, for her 'there's no point talking'. Lynn's reactions echoed the bulk of responses to the other, more conflictive, story of Susan and Harry given below.

Daily conflict

This vignette is about unequal shares of housework and childcare after work hours, a case of explosive conflict over unequal involvement with home life.

Susan and Harry have two children aged three and five. They both work full-time and Susan's mother looks after the children. They do not have any other domestic help. When they get home after work, Susan makes dinner, makes the beds, puts the laundry on, hoovers the floor, feeds the children, gives them a bath, puts them to bed, gives Harry his dinner and does the washing up. Harry goes shopping in the supermarket or shops nearby, brings the shopping home and goes to the pub until dinner time, after which he watches the telly. Susan thinks this situation is unfair because she is also tired and needs to carry on working. Do you think that Susan should ask Harry to help her to do some of the things that she has to do?

YES. What? make beds, do the laundry, hoover floors, feed the children, bathe or put the children into bed, look after his own dinner, do the washing up ...?

NO.

There was overwhelming agreement that Harry should 'help' Susan to do some housework and childcare. Some people had very strong reactions, saying that Harry was a male chauvinist and that Susan should leave him. Body language was prominent throughout my reading of the story: giggles, laughs, noddings and comments of indignation, agreement and surprise. However, after a first strong reaction there were plenty of identifications with illustrations from personal stories or stories about friends and relatives. Sometimes the personal story offered was more directly related to one of the two earlier stories, but I often felt that the emotions triggered by Susan and Harry's dilemma opened up disclosure more readily than the other vignettes.

Half of the respondents thought that Harry and Susan should share everything equally, while a quarter thought Harry should help with the children. More women than men mentioned involvement with children's activities, whereas the men tended to also mention activities of housework. It is interesting that, for women, men's help was found to be best segregated to activities related to children only. Of course, this is a story where children are very young. But more men than women included among the activities that Harry should do things like washing up, hoovering, cooking, and even making beds and doing the laundry. However, in real life only a few of those men actually did these tasks.

Should Susan ask Harry to help her?

'I think she should leave him', said Janet Seaman (H8), laughing loudly and concluding: 'Most definitely!' Her husband Daniel agreed that: 'He's got to do at least half what she's doing.'

Tracy Green (H1) said: 'Leave him.' 'Yes, she should ask him to look after the children while she's preparing the evening meal. If it doesn't work she should leave him.' However, unlike Tracy, her husband Gabriel thought that Harry should 'get more involved in meal preparation'. 'He may have a real problem about dealing with young children – to a large extent I've felt the same', he said. And we know that Gabriel avoided getting home in time to see his children before they went to bed:

> I think she should get a divorce! [*laughs*] Get rid of him! [*laughs*] What could she ask for? Oh, crikey. Make sure they've got all the technical things necessary ... to make sure they can do it as quick as possible. Try and even the workload, that's just so unfair. (Katie, H2)

These opinions mix people living in more traditional and also less traditional patterns of gender divisions in the home, and people with quite different levels of education and wealth. Tolerance for the 'unbalanced burdens' of Susan and Harry was very low across the board.

Rose Chambers (H5) said: 'They have to decide who's gonna do what ... it's very one sided. He certainly shouldn't be going to the pub while she works!' Her husband Ronald identified: 'This is similar to the problem I have. If I'm doing domestic work and Rose's not helping, even if she's done it earlier, it upsets me that she's not helping ...':

> It doesn't matter which [job]. Like us: because Deborah [her daughter] is mainly my responsibility, and Eleanor likes ironing ... she will do

more ... You can sort of work it out, but somebody does more of one and the other more of another, depending on what you like doing or what you are better at, if that works. If it doesn't then allocate him a job! (Rebecca, H17)

But, of course, there were some dissenting, more conservative views. Rosanne Goodman (H6) stated:

It's a difficult situation, isn't it? Depends on whose choice it is. She's been laid back not making the beds before going to work. For them ... maybe she needs to go out to work, maybe she drives him out of helping ...

The story evolves as follows.

Susan asks for help but Harry says that he does not know how to do these things that she has to do, and that the children want only 'mummy' at this time of day. However, Susan insists that he should hoover the floors, help with the laundry and do the washing up. But Harry does not hoover the floors well, the laundry gets stained and is not well cleaned, and the washing up is only half done. Do you think that Susan should:

(a) teach Harry to do the work well?
(b) send Harry back to the pub, leave him to watch TV, and do it all by herself?
(c) have a row with Harry and accuse him of deliberately doing his jobs badly just to prove that he does not know how to do 'housework'?

Harry does not do the work well. What should Susan do? More than half thought Susan should teach Harry to do the work well. Over half of the women and slightly more than half of the men chose this option. More women (a third) than men (a quarter) thought that Susan should have a row with Harry. More men (a quarter) than women (a sixth) thought that teaching and rowing would go together. On the whole, men seemed more understanding of Harry's situation:

A row seems more attractive, but talking is more sensible. ... it's in everybody's interests for ... everybody to be happy. They have to talk and find out why Harry isn't doing it very well. Perhaps he's not had a good relationship with his children. – It's sad. Perhaps he needs to learn – to spend more time with his children. (Ronald Chambers, H5)

I say teach but wishing to say have a row. ... It would be more sensible to teach him. Children become adjusted to the other parent. He becomes proficient. The problem is that Harry doesn't just throw in ... it brings despair and anger. (Rose Chambers)

The bit about the kids are used to the mother doing ... that's only because he's never done it with them before anyway ... If he starts getting involved with the kids, ... he's not gonna know till he starts doing it what he's good at so ... (Raj MacDonald, H19)

Children may be wanting mum to do it ... But [Harry should] start doing it ... it breaks down the patterns, the situation changes. (Marc Churchill, H16)

Old habits die hard. She'll have to talk to him on the off chance that he doesn't really know what to do. Then yes, he needs to be taught. (Richard Bartholomew, H20)

Certainly, there were alternative understandings. The full-time house-wife, Rosanne (H6), thought Susan should do nothing, but should sim-ply carry on doing her tasks. Colin Addison (H3), who did nothing at home, thought Susan should send Harry back to the pub. Gabriel Green (H1), while finding the case too complex to give an opinion, said: 'Harry doesn't want to do it. Susan will have to compromise her standards or do it herself.' Accordingly, Gabriel's wife, Tracey, also said that 'nobody does it as well as yourself. He [Harry] could be trying and not doing it deliberately. Not a lot of men, generally, know how to do these things'.

However, tolerance and teaching were not the best route for some. Why should Susan have a row? 'Because anyone who cannot do washing up, the floor ... is a liar. And he's making a bad job of it!' (Trevor Lakin, H4). Similarly, Uli Naylor (H15) said that Susan should have a row because Harry was deliberately doing things badly. 'But I imagine he wouldn't learn. It's probably a cultural thing.' Uli's wife, Jane, seemed to agree with his cultural interpretation, but her stress on 'agency' was greater. She had said in response to the first half of this vignette: 'My husband wouldn't get out of the door until everything's done ...' In relation to the second half of the story she said: 'Supervise him. Treat him like a child. Because he's a male he doesn't know how? Of course he'll learn. Uli learned!'

The story finished with the possibility of a row, but more disastrous views were offered. Interestingly, these views are from working-class individuals.

Daniel Seaman (H8) said: 'There's a divorce coming there. Mum's looked after him ... he expects the same ...' His wife, Janet, as we saw above, had told me after the first half of the story that Susan should 'most definitely' leave Harry. After the story evolved she said Susan should 'teach him':

Elizabeth: But you said she should have left him.
Janet: Yeah, but she didn't, did she? Yeah, I think she should have got out a long time ago. ... divorced him, or whatever, but she didn't. You continued then to say that she stayed.
Elizabeth: Yes, so given that she stayed, she should ...
Janet: Yeah, she should teach him and if he doesn't want to [do things], she should kick him out or whatever.

I think she should have a row with him but this wouldn't take her anywhere. He sounds like a horrible man. Get him to buy a dishwasher. By the sound of it he'll never improve. I'd not like to live with somebody like that. I'd get rid of him. (Katie, H2)

There were also plenty of identifications that took the situation very seriously:

Fancy having somebody like that ... I suppose to an extent I was a bit like that with Karen, really, I mean ... I used to do an awful lot more ... Well ... she's got to either have a row or negotiate more but she's got to persevere or leave or get him to leave! (Rebecca, H17)

Elizabeth: Do you think he will learn?
Diane (H16): Oh absolutely! ... This one I married, he was 19. He'd never lifted a plate, his mum did everything. Now he does more than most men do. He's very good around the house.

Well, it sounds like Ray this ... if I say 'Will you do the kitchen?' he thinks washing up is doing the kitchen, he doesn't think of wiping the pots and doing the floor ... (Lindsay, H12)

This story raised strong emotions. The individual reactions seemed to be consistent with gendered or personal experiences of caring for children. Also, it seems that more trust in the possibility of change was voiced by those who had experienced greater life changes themselves and who were more strongly committed to a pattern of more equal sharing in their home lives. I turn now to the dilemmas and the ethical choices implicated in the three stories.

Considering domestic dilemmas

The contexts of the stories of the vignettes were not rigidly fixed and could be interpreted and judged by each individual. The needs of the people in the stories were not taken as absolute. 'The proper thing to do' emerged from personal constructions of the situations outlined. There was no single 'true' way of reading the accounts. People filled in the gaps in their knowledge of the conditions of the characters by making their own (often undisclosed) personal assumptions. In a number of situations I was asked to fill in details about the characters' lives, but I responded with either 'I don't know' or 'what do you think?'. Appropriations of the stories with issues of projective identification were common: 'This is like my sister's', 'My friend is like that', 'This is the situation my parents were in'. Sometimes personal stories were spontaneously offered, while on other occasions I invited people to think of similar situations they could tell me about. Moral reasoning here seemed to fit in very closely with Finch's (1989) argument that moralities are not fixed, but that they change with changes in the life course, with different situations and personal circumstances.

Home lives are sites of complex moralities. Only about a fifth of partners living together, but interviewed separately, agreed about what was to be done in any one circumstance. Not one set of two people living in the same household expressed agreement in the three stories about what should be done. Only in three households were the two adults in total agreement about what should be done in two of the vignettes. In the household with the most conservative gender arrangements, Rosanne and Mike Goodman (H6) had matching views about the outcome of actions in the stories of Jane and Andrew and Emma and John. In two households at the other end of the spectrum, where the men worked from home and the women had full-time jobs out of the house, there was also agreement about two of the situations. Ray and Lindsay Wells (H12) thought the same about what should happen in the cases of both Jane and Emma, while Marc and Diane Churchill (H16) agreed about the outcomes in the Emma and Susan stories. However, many more partners expressed partial agreement about what should be done or expected. While it was rare for partners to express completely disparate views about the same case, this did happen. The significance of these multiple ethical stances lies in the increasing importance of individuals making their own choices about being moral selves, referred to in the current sociological literature (cf. Bauman, 1995). But my findings significantly contest views that there is *one* 'moral economy of the household' (cf. Silverstone, 1992). Different moral views often conflict and are negotiated.

Three issues appeared particularly important in people's considerations of how to act: (1) the need to change to achieve a balanced, 'fair' way of living together; (2) the difficulty of changing a partner's practices; and (3) the effects of parental choices on children:

(1) *The need to change* indicates an attitude of no resignation to a situation that is perceived to be damaging to the self. It implies a clear opposition to 'putting up with whatever' due to a fear of change, or a belief in anything abstract that has no implication for a better way of living the everyday. While only a few considered it better to 'keep things as they are', the dominant emphasis of responses to the three vignettes pointed to a readiness to change life circumstances. For most people it seemed possible to change one's life conditions and there was a perceived need to change in order to better a conflictive situation. The routes were various, but there were ways out to be explored. As I mentioned in the introduction to this chapter, this points against Bourdieu's sociology of agents understood to be marked by the habitus, perennially and structurally reproducing dispositions, and being particularly resilient in relation to gendered dispositions. It appears that, in line with Thevénot (2007), individuals feel induced to change in the face of situations they encounter, according to localised dynamics, judging conduct accordingly and seeking creative solutions. This is not to say, however, that properties like gender and class are insignificant, as they do affect choices. I will return to this point later on.

(2) *The difficulty of changing a partner's practices* was recognised in the different choices in response to the various scenarios. Living with John appeared to be seen as a straightforward affair. There was no particular scenario of separation and divorce in people's minds when appraising Emma and John's story. It seemed possible to find a liveable compromise in their case, while accepting that basically John's practices could not be changed much. It appears significant that this was a story where resources could be used to cushion domestic conflicts. Paid domestic help and household technologies were seen as important instruments for helping solve Emma and John's dilemmas.

On the other hand, a view prevailed that life with Harry was basically very explosive. There was a feeling that he needed a second chance, an opportunity to redeem himself, although most people did not believe he would change. However, to chuck him out before giving him another chance appeared unfair to most people. Interestingly, although the vignette itself ended with a possible scenario of a row between Susan and Harry, a number of people read the story as evolving towards separation and divorce.

The basic issues were tolerance of one's difficulties and shortcomings, but also a realistic assessment of the possibility and willingness of the other to change, while the perceived need to move on to a better situation was strong. The valuation of a way of living, mixing objects and personal relations, is implicated in the struggles in the domestic field. This evaluation of experience is central to subjectivity, as it is to social position. As Andrew Sayer (2010) argues, human beings' relation to the world is one of concern. Moral sentiments, as well as immoral sentiments, are frequently prompted by inequalities and domination. They are not merely 'feelings' or 'affect' but assessments of materially shaped social circumstances. Significantly, the responses indicated that the strongly hard-pressed life conditions of Susan and Harry demanded more immediate personal change to face conflict than the relatively better-off circumstances portrayed in the case of John and Emma.

(3) *Affecting children* appeared as a concern in very subtle and understated ways. The vignettes indicated the presence of children only in the background. However, the presence of children was felt strongly, particularly in reactions to the dilemmas of Susan and Harry. For women, Harry should help with childcare. For those men who had experiences of caring for children, and for women whose male partners had had to learn to care for children, Harry appeared incompetent and withdrawn due to a lack of experience. It was felt that he needed to 'have a go at it'. He could have a chance of succeeding.

In Jane and Andrew's story, the dealings with the video and computer brought about references to children as technically competent (John, H10) and in Emma and John's story, children appeared as needing to be taken care of after school hours (Trevor, H4). The stronger concerns about children in Susan and Harry's story was due to them being younger and the parental conflict being greater. Should people put up with more for the sake of the children? The concerns for children were an important issue in choices of everyday living. I noted in relation to the everyday routines of the families participating in the ethnographic study that childcare needs were a key marker in the structuring of adults' daily activities and of their choices about life's regularities (Chapter 3). The closer children are immersed in the adult's life, the more limited is contingent choice, and this affected more women than men. The effects of the presence of children appeared to be strongly contextual-dependent. Whereas teaching Harry was important, women's patience and possible teaching roles regarding their male partners were viewed as highly diverse. In relation to the first story, most women found that Jane had already taught Andrew and she should therefore refuse to carry on doing the task of video recording for him. Because children were the most absent in this story and most present in the last story, it seems that, unlike Jane, Susan was thought to need to be more patient and willing to have another go with Harry despite his failings, which seemed relatively worse than Andrew's. This interestingly echoes the findings by Julia Brannen *et al.* (2004) in their exploration of processes of change and continuity in family life, in particular regarding the changes of relationships along the life course and the contextual processes of negotiations (or the relevance of 'localised dynamics'; see Thévenot, 2001).

In my exploration, while material resources strongly impinged upon choices of what to do, the implications of care responsibilities were the most pressing in the moral reasoning of adults. Yet, patterns of social divisions associate with different responses. There was an evident greater propensity of interviewees from a working-class background to express less confidence in negotiations and in talk to change practices. They also placed greater emphasis on relationship breakup as the outcome of conflicts. This resonates with Lamont's (2000) findings about the senses of dignity of those in the working class being associated with personal integrity, operating within clearly delineated roles, which they saw lacking among middle-class individuals. For the former, commitments operated more strongly as an investment, reinforcing views that ethical dispositions relate to social location and material interests (Tronto, 1993: Lamont, 2000; Sayer, 2005).

Nevertheless, the prevailing moral orientations of the participants in my study pointed towards an understanding that in family relationships people are dependent, have complementary abilities and are interdependent (Griffiths, 1995). This is why, in discussing the conditions in which people interviewed made their opinions, certain difficulties were acknowledged. In the opinions expressed, there was rarely a hint of a 'correct' moral position. Although Harry's ways of living triggered judgement more readily, mostly there was no fundamental or judgemental position.

Creative adjustments to contextual and localised circumstances appear to guide collective ethical practices. Such senses of achieving compromises in family life – in a world of fluid and dynamic relationalities, yet configured by patterns based on class, gender, sexuality and so on – are further stretched with an eye to explore the involvement of technologies in the sexual life of families, which I will discuss in the next chapter.

8
Sexual Lives

When I began to explore the role of technologies in home life through my ethnographic study, I repeatedly found myself engaged with participants' sexual stories. I did not seek these; they were just part of people's lives and were embedded in their life histories and everyday businesses. This was perhaps a consequence of the method used because I explored family life by drawing on biographical and observation analyses. Perhaps this is pertinent given that sexuality is also a 'misrecognised' field in the terms of Bourdieu (with Waquant, 1992), who speaks about practices of symbolic violence resting upon the invisibility of certain social issues. The lesbian families had sexual relationship stories explicitly woven into their accounts, including issues of insemination, transitions from heterosexual to homosexual partnerships or bringing up children in unconventional families. Often the stories of the lone parents, particularly in connection with the breaking up of a relationship, were woven around a painful episode involving a sexual liaison. Heterosexuals living in partnerships offered perhaps more subtle stories within taken-for-granted 'normalised' sexual lives. Yet, some stories revealed ways in which uncommon sexualities are encompassed in otherwise ordinary occurrences of contemporary family life. My sharpest eye was directed to the interplay of technologies with ordinary sexual practices.

In this chapter I explore the links between technology and sexuality in family life to consider issues of everyday pleasures and pains and some socio-cultural patterns involved in how these two areas are related in contemporary living. Both are somewhat 'naturalised' and invisible, but are highly influential in social relations. In this technologically cultured world, relations with sexual aspects of everyday life are often embedded in practices which relate to, and take shape according to, particular technical apparatuses.

Yet, it could be said that controversies about family ethics and values, which involve sexual mores such as divorce rates, lone motherhood, same-sex marriages and unconventional sexual practices, are really discussions about how people decide what the important personal pleasures are (Phillips, 1996). But how individuals seek everyday pleasure and the resources available for this are not issues that are often engaged with in academic research. How do personal and relational choices, technological resources and identities figure in sexual practices of contemporary family lives?

Eroticism has frequently been hinted at in feminist studies of technology. New technologies, the virtual, and the body, particularly as these are linked with subjective experiences and identity, have become important areas for the exploration of experiences with the erotic (Lykke and Braidotti, 1996; Terry and Calvert, 1997; O'Toole, 1998; Juffer, 1998; Green, 2001). However, more traditional technological fields have also shown similar links, for instance, in the practice of engineering as pleasure in the domination of nature and materials (Hacker, 1989), the ways in which the language of birth and control are applied to military technology (Easlea, 1983) and references to the strong affection felt by women towards laundry and kitchen equipment (Berg, 1996). Household technology manufacturers and the media have for some time linked eroticism and housework, taking up themes of (hetero)sexuality, masculinity, femininity, pleasure and desire in marketing and advertisements. In these, the materiality involved in operating machines (to clean, cook, eat or relate) is portrayed as light and fun, while the connections objects have with 'lack' (in the Freudian sense of the 'non-possession' being the definer) or desire, beyond the desire to consume, is rarely addressed. There is a problem then in the politics of pleasure, which assigns specialness and visibility to desire as possession, danger and/or transgression, but leaves out the ordinary demands of living (Jackson, 2008: 34).

Expressing a need for researchers of sexualities to explore the ordinary effects of the material world (presumably including the technological), Ken Plummer (2008) notes that together with profound historical, social and cultural forces, economic and material influences are also at work in sexualities where inequalities of class, ethnicity, gender and age configure sexual relations.

Interestingly, however, the most salient aspect of sexuality research has been experimentation and transgressive diversity, where the 'noise' has been louder. Yet, radical biomedical technologies have come to affect ordinary sexual family life via issues of conception and reproduction. And a central feature of contemporary sexual practices, overshadowing other 'muted' mundane explorations, has been new communication technologies, which have enabled the wide circulation of intimate acts in public spaces. As computers, the Internet and mobile phones entered the home, so too did new ways of exploring sexuality and of practising 'family'. Some of these ways are subtle, reflecting the redrawing of the boundaries of the permissible, and relate to seeking or tolerating new experiences that resonate with, or do not destabilise, current patterns of 'normality'. Other practices are invisible because of the taken-for-grantedness of 'normal' sexuality linked to hegemonic masculinity and motherhood, while some others are dangerous, like the acting out of 'perverse' online desires in real life, which risk the exposure of sexual secrets to home members and unknown online eyes. Barriers, digital pseudonymity and various privacy enhancing technologies have been developed in connection with the perceived dangers of the repertoire of new sexual practices (Attwood, 2006: 79).[1]

Both 'noisy' and 'muted' sexual practices[1] enter family lives in ways that reveal the strain in contemporary social mores, as identified by Stevi Jackson and Sue Scott (2004: 244), where the 'celebration of sexual pleasure, experimentation and diversity' exists alongside 'a wariness of sexuality as a source of anxiety and revulsion'. These tensions resonate with the explorations of sexuality as an everyday aspect of social and family life in this chapter. In the next section I discuss how sexuality has come to feature in my ethnographic investigation and present three case studies to weave together technological and sexual stories. This is followed by a discussion of research on two significant fields in the 'information age' – technologies of reproduction and of communication – in relation to the case studies. The last section considers views of epochal social change linked to the impact of the 'information age' on personal lives which are challenged by my empirical findings.

Technological and sexual stories

Technologies and sexualities shape our identities, relationships and interdependencies. Yet, in relation to sexualities, the technical is not as specific or tangible as it is in other areas of everyday home life like cooking and cleaning: technologies are more subtle players in this field. This was the case in the sexual stories in my fieldwork. As noted in Chapter 1, it is only recently that researchers have begun to explore the processes of unravelling the taken-for-granted by applying more sophisticated methods of 'defamiliarisation' (Bennett and Silva, 2004). Bourdieu's notion of the 'habitus' is relevant in this context. It draws attention to how we come to be habituated to certain routines and ways of being, and thereby reproduce practices in a naturalised manner. How did sexuality become visible in this study? How did my habitus and that of the individuals I was involved with in the research enable the revelation of the sexual? In a simple answer we could see that I was 'following the connections', as prescribed by Latour (2005), but I was also attentive to a world of differences and depths.

The majority of participants in my study were heterosexuals. In research on heterosexual couples, sexuality tends to be mostly invisible because of the taken-for-grantedness of the sexual relationship. Heteronormativity is assumed to be unproblematic on an everyday basis (Richardson, 1996). But 'muted' sexual stories in heterosexual partnerships appeared in my study. Nancy Mitchell was undergoing extensive and painful fertility treatment to have a second child, even though her husband wanted no more children. Rosanne Goodman had a young baby and fantasised about the sexual prowess of foreign men, projecting onto me a knowledge of adventures that she desired. Lindsay Wells dressed in a tight, short suit skirt and was picked up by husband Ray everyday at lunch time for 'laying in' together, reported to me by both in a complicit manner that clearly implied it was a time for sex. Her exhusband lived nearby and was often around the house, and this seemed to sexually excite them both. Jane Naylor had been a lone mother at the age of 20 living in the parental home for four years before moving in with Uli, the father of her son, whom she then married. Rena Rock had never wanted babies but felt that she had not had a choice, gave birth to three sons and spent many years incapacitated by hormone disturbances while seeking treatment. A recent hysterectomy offered new hopes for both her and husband John. All of these stories came from the women. The men were more guarded in most respects, particularly when alone with me. The exception was Henry Gow, who had a very particular story to tell, which I describe below. But his story came out firstly via his wife, Lucey Lilly, though afterwards he agreed to talk to me about it.

When sexuality was revealed as relevant to my investigation, how could I interpret it? How can I know that my interpretation is 'right'? With these questions I return to a theme raised in Chapter 1 about making humans talk. In the research process I was producing answers through the framework I was setting (see Appendix 1). Yet, the diversity of the stories I found shows that accounts were not produced solely by my frameworks. The stories had similarities and complexities because people elaborated similar issues in different ways. Narrative themes of masculinity, femininity and motherhood were interwoven across individual accounts. Masculinity entered via many areas of life like the experiences of work, family roles, power in institutions and discourses, all of which acted as a frame for constructing sexuality and gender identity. Narratives related to femininity also included work, family matters, power, but frequently also motherhood, though not as an explicit concern. Motherhood was often un-thought about, although in some cases it strongly established women's credentials, providing them with an occupational identity, which was particularly relevant in the absence of realistic and satisfying alternatives, as illustrated here by Rena's case. Lesbian motherhood achieved through artificial donor insemination is, however, due to its required practical activation, highly reflexive, sitting on the verge of both 'noisy' and 'muted' sexual practices. This was the case with Josie, as shown in this chapter. The theme of masochism was another feature in the study, which I have chosen to address because of its special significance, as it appeared in this investigation, in relation to uses of new communication technologies in the home. The case enables reflection on the implications of interdependency when people living together in ordinary intimate family arrangements are involved in 'noisy' sexualities. This is the theme of Henry and Lucey's case. Masochism also relates to concerns about the body and the material. It refers to a physical bodily practice connected to sex, which in the case discussed here is also an engagement with masculinity. I will present the relevant narratives about the families before discussing more fully the implications of these sexual stories for the cultural and technological worlds of contemporary family life.

Rena and John

In Chapter 5 we learned about Rena and John. Here, it is important to remember that they lived in an East Anglian village, near Rena's parents. She worked a couple of days a week in the summer cleaning boats. She was 'never one for babies', but had mothered three sons (aged 12, 14 and 17), whom she never breastfed because she 'could not bear that sort of relationship with babies'. Every one of her sons was born quite underweight, by Caesarean section, the first with a cleft palate, which 'freaked her out', after which she 'cried for days'. She was clinically depressed during the second pregnancy, being depressed again during the third. She had difficult menstrual cycles, feeling well just for a week every month, until she had a hysterectomy, when she then began to feel 'lovely'. Five months after her hysterectomy operation she enrolled on a month's return to work course in customer service, which included computer skills. When I first saw Rena it had been four weeks since the course had ended.

John described himself as 'just an ordinary, honest, basic person'. He had a great sense of family belonging, of his duty towards it and of self-sufficiency. 'I made my house with my family and with my family I keep it up.' He was 'totally in charge of all day to day operations' and 'about 350 people' of the company he worked for, his employer for 25 years. He often spent weeks away from home and an odd week or so at home, thus needing home computer and Internet access. He had 'three fathers altogether' and had never met his biological father – the second, his 'stepfather really', from the ages of eight to 31, was an engineering manager at General Motors in Luton, where his mother also worked at the time, in the office. The third, and his mother's current husband, was not a significant person to John. He described himself as a 'sole support child', and 'very independent'. His brother, three years younger than him, was also described as highly independent and was at the time a top finance executive in a large company abroad. After finishing school, John went to 'the Forces', training for three years, then had a job for seven years, before moving to his current position. He married when he was 30. The two major changes in his life were moving 'from being an ordinary worker to being the person in charge and getting the best out of people' and 'watching the children grow up, achieving things'. When John was at home he did a lot of house maintenance and building work. He also helped with domestic chores.

I witnessed revealing episodes of the dynamics of their technology use. A new computer was installed in the lounge, and the Internet connection was new when I first visited. The five family members each had an email address. Rena was working on her family tree. They had big arguments about using the new computer and the Internet. I, embarrassingly, witnessed a couple of these. On both occasions the father's technical non-competence was pointed out by the boys and Rena. He said he did not care because 'the girls in the office' could tell him what to do.

The family's economic position and material life was perceived as good. They owned their house with no mortgage. The car was one of the perks of John's job. A very important element in their coping strategies was their household technologies. They were strategic for the subsistence and maintenance of their physical bodies and lifestyle. The use and care of the technologies for housework stood out as important both because of how they were introduced and used in the home and of how they were identified with the use and care of Rena's body. As I remarked in Chapter 6, the machines were overused, not always working properly, they were left not functioning for periods of time, the activities they performed were accomplished with lower quality substitutes, and the home had an accumulation of useless things ('I don't know what's in there', said Rena, commenting about the contents of the freezer. Bath towels were said not to have been seen for months). Rena felt overwhelmed by piles of laundry waiting to be ironed while demanding of herself that she iron everything in conformity to the traditional femininity to which she subscribed.

Bringing up children did not seem to have been a problem in either Rena's or John's narratives. The boys appeared 'self-supported', like their father had been as a boy, and the parents seemed to expect this of them. The plentiful access to individual interactive technologies reinforced the boys' self-reliance. They each had a Nintendo, TV, mobile phone and email address, and the growing intake of computers in the home would soon enable individual ownership of these too. This was one of the first families engaged in my ethnographic study. In 1999 only a quarter of households owned a personal computer, while Internet access and email use was limited to less than ten per cent of the population. Mobile phones were then a great novelty, unlike the nearly universal commodity they have become for British children in the first decade of the twenty-first century.

Beyond particular links with technology, gendered masculine and feminine cultures pervaded the performances of the individuals involved. For instance, a big difference between John and Rena was that whereas Rena formally embraced the prescribed feminine roles of housewife and mother, she lived these as adversarial. John, on the other hand, was untroubled by the cultural models of masculinity. He grew up as a self-reliant individual and his past in 'the Forces' helped him achieve a good image of himself. His job was a typically male occupation. He worked away from home with flexible hours because he was on call 24 hours a day, even in his weeks spent at home. His job required the support structure that was traditionally dependent on a wife. Rena was not able to do that on her own but the kin network, particularly her mother, supported family life during Rena's frequent illnesses. Her mother visited daily and dealt with Rena's kitchen with great familiarity, as I observed. The masculine culture that benefited John was also reflected in the independence of the sons, who appeared to cope generally well with their mother's withdrawals. The three of them knew 'how to cook', often sorting themselves out with microwave foods. They knew how to operate the washing machine, the oldest did his own laundry, and not one of them mentioned the mother's health as a problem. Rena's low level of support appeared to have been enough for them to get by with.

Rena suffered from the definition of women in terms of sex and reproduction, which identifies the female body as the 'other', essentially different. Rena's working-class origins were also reflected in her lack of proper work qualifications. She had a low educational grade and low self-esteem. With low cultural capital, she felt actually 'out of place' in life. Her story shows great conflict with the culturally assigned traditional feminine roles of wife and mother. She had had babies she did not want, suffered huge health problems related to pregnancies and had lived with depression and a lack of energy. She had some hope for a happier life following her recent hysterectomy. It appears that Rena suffered from the symbolic violence attached to the prevailing gender order which narrowed her options in life. Interestingly, her conflicts with her assigned attributions indicate ambiguities in relation to the determinations of her habitus that directed her to pursue the roles of wife and mother in a taken-for-granted manner. While in practice she undertook these roles, she did so with unease, which was evident in her suffering body and in the emotions of anxiety and revulsion she had displayed towards her motherhood. Rena's resistance to her assigned habitus stresses the need to recognise ambivalences in the ways women and men occupy gendered positions aligned with particular femininities and masculinities, an aspect overlooked by Bourdieu (see Silva, 2005 and 2006).

Josie

In Chapter 2 we met Josie Barker (42). She lived with her two young children Michael (11) and Cassie (4) in south London. Josie was white but Michael and Cassie were ethnically mixed African-Caribbean because of the ethnic background of the co-parent, Nadia. Josie and Nadia had never lived together, although they had been partners for 13 years. Both children were conceived by artificial insemination. Josie said she had wanted to have a child before she had met Nadia:

> I was exploring avenues and ... I was certainly up for doing it as a single parent and – but ye know, none of the avenues were very easy. And at that stage I was looking for a known donor – I mean when I got involved with Nadia she em – there was quite a big ... factor in lots of different ways. One: because she'd raised three children on her own so you know, without a known, without ye know known men around erm – so that answered a lot of my questions, ye know, because I met her children as adults, about sort of the effects on children of that. And – she felt very strongly that – em if we were doing it as a partnership that we didn't actually want a third person as well so – we went,

so we used a clinic so ... yeah, so Michael, ... [*talking quietly*] – and originally I sort of didn't look beyond having one child but it just didn't go away! Erm ... and in the end I just got so fed up with it still going round and round and round in my head that I just thought – oh I'll just phone the clinic and – that was the slippery slope that ended in Cassie, which I'm very pleased about. But although like sometimes I get a bit broody you know, this is definitely it!

The stories of the pregnancies were inserted in institutional changes concerning the human fertilisation and embryology legal apparatus in the UK. Josie said 'Michael was a BPAS baby' to indicate that he was conceived under the British Pregnancy Advisory Service, a National Health Service (NHS) branch which held sperm samples and carried out Intrauterine Insemination (IUI) at the time. She recounted: 'I went to PAS first and having taken all my money they then said they hadn't got any black donors, they've (the children) both got black donors – em – so – I went to BPAS. ... then BPAS stopped and so yes, I went to PAS again for Cassie and then they stopped but – in fact they stopped more or less as I fell pregnant.'[2]

Josie explained that it was a definite decision to have a black donor because:

Nadia's African Caribbean – and I – I mean ye know, she hardly had to twist my arm, I mean ye know the sort of rational and political reasons and then there's irrational stuff ... 'cos I realise now how ... ye know, what muddled thinking it was but I do remember really having this sort of clear emotional thought: God, if I had a white child and it turned out to be racist ye know, oh my god that would be so ye know, so awful, but –

Josie discussed the 'moral dilemmas' of the situations of the vignettes in my study (see Chapter 7) – all relating to heterosexual partnerships – making connections with her lesbian experience and that of her lesbian friends. She liked not living with a partner and enjoyed having the children and her own life. She felt quite stressed at times but thought that overall it was very good. Josie had close connections with two other lesbian families with children of a similar age as hers, and this constituted great support for her in terms of everyday needs and emergencies:

I've got two sort of families that I feel sort of, ye know, we're involved with each other – and one lot I met through Michael, through the

midwife that delivered Michael because she delivered their first son
Jack the day before she delivered Michael, – she couldn't believe that
she'd delivered two lesbian babies you know, one day after the
other – to her knowledge and her career to that point, she'd never
delivered another lesbian baby. So that's a sort of friendship that's
been going for nearly twelve years.

... then with Cassie, ye know, it happened again – I mean it was
slightly less chancy this time 'cos we both went to a very lesbian
friendly midwifery practice – but – yeah, she's got a very close friend,
... and I sort of – hang on out a lot with her mums, very important
to me.

Josie had experienced a different and less common form of sexuality,
femininity and motherhood. She had university education, a middle-class
background (her father was a trained physicist working as a computer
salesman and her mother was a teacher), felt professional worth and
had developed her connection and senses of belonging to a supportive
community of lesbian mothers. In contrast to Rena, her motherhood
experience had been consciously sought and desired. She was less
secure financially than Rena, but she did not seem to feel the poverty
as an entrapment.[3] On the contrary, she appeared full of life, on the
move, making choices, making connections and arranging her own
affairs. Her children were a major responsibility in her life and were
central to how she organised her routines and overall resources. The
technologies in the home were just an effective part of the household.
They would be bigger and better if there were more space and money
available, as they would live better if there were more resources, and
Josie was seeking those, but she was aware of her limited economic
resources. Yet, she did not feel lacking and did not complain or resent
her condition. Quite the opposite in fact – she appeared to enjoy her
life and her choices, which were crucially shaped by contemporary
technological resources.

Lucey and Henry

Unlike for Rena, John and Josie, I have not given a fuller presentation of
Lucey and Henry in previous chapters. Lucey Lilly (50) was a solicitor,
white, foreign but had been living in the UK for over 20 years. Henry
Gow (53), a British orthopaedic consultant, was her husband. They
lived in the north of England, and had a 14-year-old son, Stuart, who
was being educated at a private school. They had both travelled a lot
and had had jobs abroad. They lived in a detached stone Victorian

house with a large garden. All sorts of home technologies were available in the house, including cable, one computer and three laptops, and in 2002, as early users, they also had broadband and a wireless Internet connection.[4] Lucey's family lived in Canada, while Henry's lived in London.

Lucey's mother had been a teacher, while her father had been a businessman and politician; the family had been prominent in her town of origin. She had two older siblings. She saw her family every year and was closely connected with her parents and siblings via travels, letters, emails and phone calls. She had had a number of sexual partnerships, including with a woman; this was her second marriage. She had wanted to have a child by the time she met Henry, in New York, while she was on an internship and he was on a medical residence course. When her son was born, it meant slowing down professionally. They moved to England soon after becoming parents. Lucey qualified to work in the UK and secured a post. She felt in conflict with 'the British prevailing ideology of stay-at-home-mothers of the early 1990s', which she felt 'antagonised' working-mothers of pre-schoolers. Her professional identity was important to her; she had held powerful posts in Canadian local government, but she greatly enjoyed being a mother to her son. She and Henry had demanding jobs but had tried to share housework and parenting on a 'fair' basis. Lucey was quite conscious that her fair share idea demanded a nearly 50-50 split. This was difficult to achieve and created lots of tension. A 'cleaning lady was essential to save the marriage', she said. Technology was also important in this respect: they owned a dishwasher, a tumble dryer, a washing machine, an Aga cooker and various other good quality gadgets and machines. Their strategy towards household appliances resembled that of the Churchills (see Chapter 5).

Henry's mother was a housewife; his father had been a GP. He had two younger sisters. He rarely saw his family, occasionally exchanging an email with a sibling or phoning his mother. These contacts involved either birthdays or money matters. His work was very central to his routines. However, Lucey and Stuart seemed to be the real core of his life.

About one year after I first contacted this family to become part of my study, and after having spent time in their home and interviewing both Lucey and Henry, what Lucey calls a 'tragedy' occurred. I became aware of this occurrence over one year after it had happened. On an entirely casual occurrence, Lucey had found on Henry's printer, in his study, a page describing a sadomasochist (SM) session. Confronting Henry, she found that he had been, in his words, 'an active masochist' for as long as he could remember. His words to me about this were as follows:

> I have known myself as a masochist all my life. As a boy I'd hurt myself in unimaginable ways. Of course I didn't know what to call this until my twenties. I simply thought I was weird gay. I discovered I liked women in my early twenties and then began to have SM sessions with dominatrices.

At the time of the discovery, Lucey and Henry had been together for over 20 years. His job had often taken him away to meetings and conferences. He had timed his SM sessions so that his body would be 'healed' by the time he came back home. He had always been so careful that he was utterly surprised at Lucey's discovery. 'I love Lucey and wouldn't know what to do without her. She was *never* meant to know about this', he said. Lucey saw the printout of an SM session description, Polaroid photographs of scenes of torture, a web film of a session the printout related to, and various pornographic websites listed under Henry's 'favorites' on his Internet browser. These sites, said Lucey, were 'racist, sexist, offensive, grotesque, unimaginable'. She asked herself how she could not have seen this for so long and she reflected:

> He pays a stranger woman semi-dressed in a kinky outfit to bondage, whip and torture him in a weird candle lit cave-like-room. It's awful to know he gets pleasure from this! I know he has a very particular childhood story that could explain that but I cannot understand how an intelligent and sensitive man could get hooked into something like this.

On the other hand, Henry asked: 'How could she have discovered this so well-kept secret of a life time now?' Their world became entirely different, bewildering and tragic for both of them. He had never trusted her enough to share any of his 'other' sexual activities with Lucey. How did this affect their relationship?

Writing about secrets and lies, Carol Smart (2007) says that secrets and the creation of 'fiction' may be felt to be necessary for the preservation of relationships. In these cases, secrets matter more than truths. Interestingly, the revelation of Henry's secret had been an act of everyday forgetfulness of a piece of paper with written information on it. It was banal and yet life-changing. The secret was connected to a closeted sexual practice and to a specific contemporary development in information technology.

Subjectivity and technologies of reproduction and communication

Assessing changes in 'the information age', Manuel Castells (1997a: 235) claims that 'there is a sexual revolution in the making' characterised by 'the de-linking of marriage, family, heterosexuality, and sexual expression (or desire ...)'. Building on ideas developed by Anthony Giddens (1992) and abundant US statistical references by Edward Laumann *et al.* (1994), he claims that 'a story of consumerism, experimentation, and eroticism in the process of deserting conjugal bedrooms, and still watching for new modes of expression' is revealed (Castells, 1997b: 239).

How do these broad patterns relate to the cases in this chapter? It is relevant to consider changes in the two technological fields involved most directly in the stories: reproduction and communication.

New reproductive technologies

Robert Connell (2002: 47) stresses that bodies are both objects and agents of social practice which form and also reform social structures and personal trajectories. We saw some of this in operation in the cases of Rena and Diane discussed in Chapter 5. But the agency of bodies and their relations with social and personal stories is most evident in the case of reproduction, be it technologically assisted or otherwise. The materiality of bodies matters in their capacities to give birth, milk and pleasure, and many social processes are informed by specific bodily matters like practices of childbirth, infant care and sexuality. All of these are implicated in gender categories. Yet, broad categories of body and gender link to biological and cultural processes in ways that are fundamentally challenged by technologies. This is not so much because people's relationships change, but because change is embraced and recognised by individuals through known and experienced processes which seek to stabilise social life in processes of change. To unravel this I want to return to Josie's case.

The technological process Josie used for insemination is the most common one involving the insertion of prepared sperm into the cervix to be placed high inside the uterus using a catheter. The procedure takes place during the woman's ovulation period. It is a painless procedure which for Josie took no more than half an hour and cost less than £1,000. The cost at the time included buying the sperm from the bank of anonymous donors. She was successful in getting pregnant following her first attempts.

Artificial insemination encompasses a large field of practices, where IUI and In Vitro Fertilisation (IVF) procedures predominate.

Normalisation derives both from its long history and widespread use. The first test tube baby was born in Manchester, UK, in 1978, employing one of the more complex practices of artificial insemination, IVF. Since then approximately 2,000,000 IVF babies have been born worldwide over a period of 30 years. The London Women's Clinic claims that since 1998 it has helped 'more than 2,000 single women and lesbian couples in their desire to have healthy babies' (LWC, 2009). These numbers indicate artificial insemination as a widespread normalised practice, but the deployment of artificial means to achieve procreation carries on challenging ideas about relationships.

This was the case where lesbian 'reproduction' took place. Josie and Nadia created offspring and became parents not by virtue of biological developments alone but by social recognition of their cultural parenthood. The genetic substances involved in the conceptions of Michael and Cassie became somewhat irrelevant. As pointed out by Jeanette Edwards *et al.* (1999), reproductive technology creates a separation between the social and genetic parent. This is not dissimilar to what happens in cases of adoption, leading me to agree with Natalia Gerodetti and Véronique Mottier (2009: 151), who observe that much of the field of reproductive technology continues to be played out in terms similar to those of 'old' technologies. I add, following Edwards (1999: 64), that the new technologies are here modelling 'new possibilities on old facts' as social arrangements gain prominence over the blood tie, yet re-script the family as a long-standing old relationship.

If in a biological sense relatedness exists without it being known – the anonymous sperm donors have a relation to the children – the technologically solved fertility 'problem' in this story is not purely biological. It is sexual and it is cultural, as shown in Nadia and Josie's conclusion that it would have been inappropriate for Josie to couple with a man to procreate. The practice followed breaks the link between sex and reproduction. The technology here mediates the moral question of how many embodied individuals would be involved in the couple's relationship of procreation and of what sort of ethnic material would best fit the envisaged family relationships.

The story gives credit to Marilyn Strathern's (1999) reading of Janet Finch's (1989) appraisal of family obligation and social change in the UK as depicting kin relationships in terms of a special connection between social relations and natural relations. The connection exists as a cultural reality drawn upon in people's social arrangements. Translating this into assessments of relations arising from technologies of procreation, obligations and ties becomes not just about how human beings come to exist, but also about how relationships emerge. The active desire for a child of a particular ethnicity was central to the technological practice followed by Josie. This was the opposite of the 'natural' process resulting in Rena's motherhood, which appeared conflictive and somewhat damaging to Rena's personhood.

In the words of Sarah Franklin (1999: 161), new reproductive technologies create new conceptions of relationships in the procreative sense and also new cultural and imaginative ways of thinking about these relations. New reproductive technologies challenge the understanding of the old dilemmas of relations between biology and culture in a technologically shaped world, and the relations between people committed to each other. It is in this sense that Erica Haimes (1998) shows that the donor-inseminated child is the product of technology and of social relations and social processes. The workings of culture can be made evident through its technologies, as culture itself is a source of innovation required for technical developments. There would actually be no lesbian parenting if the artificial insemination of a lesbian woman were not accompanied by the cultural recognition of homosexual partnership and parenting. The culture that enabled Josie to achieve her desire to be a mother was technologically assisted and the localised dynamics (cf. Thévenot, 2001) that made her feel comfortable in doing so embraced an innovative lifestyle possible in this technological world.

New communication technologies

The revelation of the closeted Henry to Lucey arose from online activity, the printout emerging from an Internet connection. Cookies constructed Henry's digital profile on the web.[5] Henry's construction of his pseudo-identity on the Internet was flawed, both because of his misplaced sense of security, which was impaired by a banal act of forgetfulness, and the myriad of what he claimed to be 'false' information that was attached to his pseudo-identity (or closeted self). Although careful with the construction of his digital self, Henry's care with this construction was ultimately useless because it was his embodied self who made the act of 'forgetting'. But there is another way in which his sense of the safety of his digital identity seems to have exacerbated the problem, creating a self for Henry which he did not identify with, but which was one further dimension of the unknown Henry that Lucey was confronted with. The cookies' targeting of Henry's machine profiled him as a closeted active masochist, who he acknowledged himself to be, and hard-porn user, which he claimed not to be. While this technology zoned, hid and protected Henry, there was no way of knowing who was sending him information, who he was relating to or who was watching what he was doing. Henry subscribed to mailing lists and erotic SM film sites. These involved fantasies, photos and the organisation of actual meetings. He used Hotmail for his emailing and Google for his searches. He chose pseudonyms to view erotic photos, to visit websites aimed at those who shared his sexual tastes. He sought information on clubs and activities in the cities to which he made business trips. He visited chat rooms, chatted with those compatible with his interests and shared his email address. He believed nothing could be intercepted or linked to his Internet Service Provider (ISP), to his name to any other email addresses he had. The web film of his sorry revelation was sold online. Although he admitted doing all of these, he said he did not identify with the hard-porn material linked to his computer 'favourite' sites, which Lucey found so horrifying. Lucey went through 'frantic searches', overcoming Henry's online protections thanks to her own technological expertise, which was better than his. In real life it was left to Henry to convince Lucey that the limits of his transgression did not reach the most horrifying levels of what she had been confronted with.[6]

Controversies inform the debate about the role of new communication technologies in personal life. Sherry Turkle (1995) shows enthusiasm for the potential of the Internet to help individuals express and explore new aspects of the self, play with their identities and create new ones. The fact that significant segments of many people's lives are to an increasing extent lived in virtual reality increases the experience of the self as decentred, multiple and fluid, she argues. Judy Wajcman (2002) further stresses that the new possibilities of sociality in cyberspace create new communities which share the need for communication and belonging without the imposition of traditional constraints, duties and responsibilities. However, she warns that this is reflected in novel forms of cyberspace technoculture, often characterised by violence and pornography.

This is a culture facing the perils of the 'information age', as noted by Castells (1997). The wide varieties of porn emerging in the last 20 years in the wake of technological innovations have sought different audiences (O'Toole, 1998; Juffer, 1998 and 2004), turning imagery and language that would have been classed as pornographic not very long ago into part and parcel of popular culture (Attwood, 2002). The potential of porn delivery via new technology has been causing some alarm, because pornography has been associated with a whole range of social anxieties like sexism, violence against women and children, xenophobia, neglect of the family and commodification of pleasure (Williams, 1989). Less threateningly, pornography has also been regarded as a kind of personal therapy and self-expression (Simpson, 1994) or even as a convenient vehicle for the easy and unconstrained pleasure of pressured and busy female lives (Juffer, 2004). In either the alarming or tamed scenarios, pornography implies a kind of autosexuality, relocated away from relationships with others, family, community or reproduction. It is free, for some, from the 'mess' and 'inconvenience' of relating with, or having sex with, other people. These practices seek to avoid the inter-personal control of social life implicated in the sharing of emotions (Phillips, 2002; Attwood, 2009).

However, the leak out of the cyber closet is not simply a technological matter, as Henry's experience shows. Embodied culture and personal relations are deeply implicated in the technological. The ways in which Henry's deceit could be borne by Lucey at a personal level is highly relevant, but it is also culturally relevant that she found that there were no social 'structures of feelings' (see Williams, 1977) which could give recognition to what she was going through. 'There was a time of just bewilderment, when I wished Henry had died and I could tell the world of my loss and pain', she said. How this revelation and process affected Henry and Lucey's relationship is again a personal, relational, but socially and culturally relevant issue. Increasingly, contemporary erotic life has been 'noisily' shown in the public media and captured on domestic videos. There is satisfaction for some in the act of being photographed, video recorded and placed on the web. Secrets of private life that formerly one would give anything to conceal are now actively posted on websites and television. Whereas Henry wished to reveal a pseudo-self with his photographs and website film, the technological culture where pseudo-selves exist is in fact real, made of embodied beings who feel *real* pain.

Much is new and a lot is old here. How far do new technologies desta-bilise current patterns of normality? Communication technologies like computers and mobile phones have been studied for their abilities to change social and personal life and yet studies have consistently returned to various stabilisation scenarios. For example, Rich Ling (2004) ends his book on mobile phones with a series of possible vignettes about his children's future lives and their relations with him, expressing a desire for unhindered communication, for a rich personal relationship between them. Similarly, Elaine Lally (2002) explores the uses of the computer in households as a medium for imagining the future where teleworking, home offices, Internet communities and so on coexist in ordinary home life. However, novelty is constrained since, as Lally states, we appropriate computers and they appropriate us in return. Relational appropriation thus connects the self and the environment, our subjectivity and the materiality of everyday life.

The technological and the sexual in research and culture

To address the links between personal lives and the contemporary technology-drenched culture, some theories of epochal change that have lately emerged have referred specifically to the body and sexuality. Castells' portrayal of social change offers support to Giddens' proposition (1992: 175) that 'sexuality becomes the property of the individual', to offer a futuristic scenario where 'by assuming the body as identity principle, away from the institutions of patriarchalism, the multiplicity of sexual expressions empowers the individual in the arduous (re)construction of her/his personality'. But before this is achieved, so Castells' story goes, 'the breakdown of the patriarchal family ... is indeed giving way simultaneously to the normalization of sexuality (porno movies in prime-time television), and to the spread of senseless violence in society through the back alleys of ... perversion'. He sees in networked society individual anxiety and social violence escalating until 'a reconstructed egalitarian family better suited to free women, informed children, and uncertain men' is formed (Castells, 1997b: 241–2).

Sophisticated and perceptive as Castells' and Giddens' knowledge of social trends is, the portrayal they present of epochal change appears to falter in the face of the individual life stories in my ethnography. Transformed gender and sexuality lie at the centre of socio-cultural change envisaged by Giddens and Castells, and technology propels some of their material transformation. In the case studies of my investigation, sex is everywhere and it can be painful. However, most importantly, this is the case in relational ways and sexuality is not individualised in these scenarios. It was often in episodes of emotional pain that the stories of the participants in my study revealed them as sexual beings. The narratives moved the private to the public, from the sometimes dark closets to where it could be 'seen'. The ways in which sexual bodies and beings became visible in my study relied on the binding of their stories with a connection between the material and the culturally elusive ties that give meaning to personhood and relationships. I have sought to destabilise the connection between the material as visible and the cultural as invisible by means of these detailed household case studies exploring the intermeshing of the technological with the personal (and the sexual). I have drawn from various sources to argue that the material exceeds what we can see, as well as that to see is not necessarily to know.

In doing this, both the fields of technology studies and sexualities were somewhat taken to task.

Wendy Faulkner (2002: 93) says the world of technology is made to feel remote and overwhelmingly powerful as the cultural images of technology remain associated with large technological systems connected with powerful institutions. Other technologies that we meet on a daily basis, relate to, and make some sense of, are not readily identified as 'technology'. This can be the case for the artificial insemination undertaken by Josie, the machines to aid housework brought into Rena's home following her pregnancies and related health problems, or Henry's printer. To exclude these would lead to blindness about significant aspects of the technological world and of personal lives. Yet, explorations of the significance of sexuality in everyday life are important when technologies in the home so strongly connect personal and collective or public stories. In pursuing connections between technologies and sexual lives we see the revelation of inconsistencies and paradoxes in the family stories. But technology is just one of the ingredients in these stories, albeit occupying a central role in contemporary culture.

Conclusions

In the descriptions of everyday events of individual lives in this book, the technological materiality of the contemporary world is prominent. Yet the technological material is so embedded in the relationality of social life, absorbed into the interstices of ordinary doing and getting about, that its full significance goes unnoticed. To see the marks of technologies on relationships, the ways in which our relationships with these materials are vital to us, the book considered the personal, the relational and the material in everyday life in terms of broader and long-standing social issues. Grand sociological questions and ordinary living/lives were brought together theoretically and empirically.

I started this book with questions about the apparent lack of effect of profound technological changes on patterns of care and ways of living everyday life. I remarked on the reconfiguration and persistence of social divisions, and in particular those of gender, even amidst increasing diversity of ways of living. I enquired about the role of power in shaping practices of change and about the connections between personal ways of relating and those of the technological order. While it is clear from the material assembled in this book that contemporary home life practices are built around an active engagement with various technologies, the technologies are themselves resources which construct social positions (and social divisions). Focusing on everyday domestic practices, I addressed broad sociological questions, conducting a conversation between social theory in various academic fields and the research material of my investigation. I have aimed to engage the reader with the lived experiences that inform the analysis and with the social processes underway over time. I accept that readers make their own choices about what and how to know, engaging more or less with the social theories, the narratives of everyday life or the archival material (including statistics). My concern with how practices of everyday home life are constitutive of – and constituted in – broader social life also extends to my reflexive practices of undertaking research both through my engagement with the generation of material and through the analysis itself.

The development of my approach in this book reflects on some models of seeing the social from technology and science studies, feminist perspectives, sociology and cultural studies. I have engaged centrally with the works of Pierre Bourdieu and Bruno Latour and their critics. In applying the frameworks of Bourdieu and Latour I noted both their productive aspects to developing a theoretical and empirical approach to bring the unnoticed to bear on larger social theories and also the ways in which they fail to grasp significant concerns in the fields of home, family and everyday life.

Given the prominence of my use of Bourdieu's and Latour's thinking, it is important to revisit their contribution to the discussion in this book. Firstly, they both recognise that specific technological assemblies are evidence of social life and that claiming visibility for technologies as resources for living implies defining them in cultural terms. Significantly, the operations of culture and the operations of technologies actively make, act and transform the social world as they assemble individuals and things together. Both Bourdieu and Latour work with this view of the active operations of the technological and the cultural fields in their concerns with the material processes through which cultural entities – like aesthetics, tastes for objects, ways of doing things and abilities to possess and engage with various materials – are produced in diverse cultural settings. Bourdieu's and Latour's views about what makes the world go round, the constituents of the social world and how to grasp their workings are, however, different from each other. While informed by their knowledge and adopting various elements of their frameworks in this book, I have been selective in adopting their approaches in my investigation. My focus on family life challenges some of their theories. My empirical engagement with 'the family' – as practice, interdependent relationality and setting – also destabilises other accounts of the operations of technologies in contemporary social life based on visions of epochal social change. I engage with these issues in this conclusion, focusing on three main aspects: (1) the mutual influences of broader social patterns and ordinary living in the context of immediate occurrences and long-term processes of social change; (2) the relations between the material and the social considering 'things' as resources in social life; and (3) the resources involved in relationality both concerning concrete practices of living and issues of empirical knowledge.

Before focusing on these aspects, however, it is useful to reflect further on my use of the work of Bourdieu and Latour.

Connections and relationalities in Bourdieu and Latour

I have applied Latour's approach (especially 2005) in my use of such research practices as descriptions, 'following the actor' and 'making connections' without pre-distinguishing 'sizes of agents'. The scales involved in particular individual and household stories varied, as they also did in cases of particular technological developments of appliances and related activities. But I have not adopted Latour's 'flat ontology' and reliance on decontextualised performances. Instead, in line with Bourdieu, I have stressed the role of social properties in both materialities and sociabilities, even though I was not always able to establish precise links between properties and the patterns of assemblies and practices that I observed. I have assumed, and found corroborating evidence in my research, that there are reasons for things going together (the consumption of particular technologies and an individual profile, for example) which are conditioned by position in social space. These reasons did not change according to each context but they obeyed certain logics of repetition across social fields. Contrary to Latour (2005), such patterns affect the relationality of the social in predictable ways.

Despite aligning with Bourdieu in finding a complex and patterned world with peaks and troughs, my approach contrasts with Bourdieu's in other respects. I have placed greater stress than he does on the role of personal agency. There are many disputes about Bourdieu's notion of the habitus. This includes the question of how far agency can disrupt the habitus, a debate that goes on even within his own oeuvre. In *Distinction* (Bourdieu, 1984) the habitus of individuals and the collective habitus of social classes is defined as being a unique, unified and coherent set of tastes determined by social origin and strongly impinging upon social trajectory. Various criticisms of this notion have emerged, including that by Bernard Lahire (2004), who disputes Bourdieu's notion on the basis of empirical material. Lahire shows that most people have 'dissonant' cultural tastes in relation to their class boundaries and that there is no overall homogeneity in class groups. With this he challenges the social determinations of class of origin, stressing the need to recognise more heterogeneous individuals and the collective formation of personhood. This is an idea that resonates with my studies and is mostly evident in relation to gender where conflictual subject positions or multiple subjectivities coexist and are manifested in practices. The comparison of the cases of Josie Barker and Rose Chambers in Chapter 2 is an illustration of this individual dissonance within similar class positions. However, the contrasting cases of Rena Rock and Diane Churchill in Chapter 5 would more readily fit a notion of homogeneity between the habitus and class as identified in their different practices. The fact that homogeneity does not always occur and that the presumed matching of habitus and social position is sometimes challenged needs to be considered to account for contemporary practices such as those described in this book.

For instance, gender-related properties required for doing femininity and masculinity are dynamic and these are nowadays no longer stuck to the person. Gendered properties are considerably fluid, organised in culture externally to individuals who are increasingly able to 'pick and mix' from different properties of gender. Lisa Adkins (2005) has examined their operation in the labour market and I extend this examination to domestic life.

In the terms outlined by Laurent Thévenot (2001) regarding a revision of Bourdieu's theory of practice, people make creative adaptations to localised dynamics shaping individual and collective social practices. For example, the skills to mother and to manage the home transcend a feminine domain and are not a property of the female procreating person. Men also 'mother' and manage homes. Creative adaptations of time use and spatial location of everyday activities have led men to develop 'feminine' roles and women to take on male identified gendered properties, both in the home and in the labour market. Heterosexual families do this, as the Chambers (H5) in Chapter 2, the Websters (H13) in Chapter 3 and the Churchills (H16) in Chapter 5 illustrate so well. Interestingly, these households are in the professional-executive group for whom practices appear more flexible because personal resources to deal with available social roles are greater. But sexuality is relevant to subverting the operation of gendered practices, as the cases of the lesbian households, Turner-Hill (H17) in Chapter 4, James-Cox (H23) in Chapter 5 and Barker (H22) in Chapters 2 and 8, show. Although creative adaptations do not appear as a prominent pattern, they are very noticeable and indicate ways in which personal and home-based dynamics inform collective social practices, even if they occur in a process of 'lagged adaptation' (cf. Gershuny *et al.*, 1994; also McNay, 2000). The transformations of gender-related properties do not, of course, happen in a vacuum but are context-oriented. For instance, as I showed in Chapter 4, the cooking practices of women and men have changed as part of a 'technological nexus', and class divisions are apparent in kitchen consumption practices through diverse kinds of cooking which convey social distinctions. Issues of personal resources, including technical and emotional capitals and moral reasoning, are relevant for social position, as discussed in Chapters 6 and 7. Both subjectivity and competences are here linked, in that even when one is born with certain resources, these do not appear ever to be fully given, having to be built, modelled or, moreover, subscribed to (often as a local and provisional choice). I consider these ideas further in the last section here.

As I moved through with the analysis of the technologies and practices in everyday life, the individual accounts sometimes clashed with social patterns, while at other times they fed into one another or fitted in some respects and challenged others, resulting in an interesting messy mosaic requiring some interpretation to make sense of things. How do dimensions of the patterned social and everyday relate? How do temporal dimensions of the 'now' and the long term interact?

Social patterns and ordinary living: immediate occurrences and long-term change

Social changes reflect minute experiences in everyday life. They reflect scale (the self and social patterns), speed (slow or accelerated) and noticeability (the capacity to see the ordinary). How can we construct models for noticing change in a relationality which takes into account technology and culture as resources impinging on relations and producing relations, including those of 'family'?

I have discussed how changes in family life and technologies have been linked to a sense of speedy current social transformations, which may lead to breakdown or threats to a sense of home, disrupt feelings of belonging or interfere with 'normal' workings of families and households. I have argued, and then demonstrated with various case studies, that there are some crucial epistemological problems in analyses that suggest that current world transformations are threatening a known way of making home lives. Firstly, homes and families are not isolated from transformations in the world at large. I note the interplay between choices of ways of living and societal changes, asserting that family and home are *not* dependent and powerless spheres. Secondly, home and family forms have always been transient, both in history and over the individual life course. Because home living is done within a culture, it is also embedded in historical cultural changes. Since individual life courses demand arrangements to fit with personal biological developments and relational needs and choices, the ways homes are made also vary as personal biographies develop historically and over lifetimes.

The types of families with dependent children which feature in my ethnographic study fit with a particular phase of most individuals' life course. Yet, in their diversity we see them breaking away from historically traditional social arrangements. The greater diversity of life choices and home living goes in tandem with the materiality in the social: from features of housing to ways of cooking and cleaning, the moral choices and ways of relating.

I have noted that the way in which technology affects the meanings of everyday life, relationships and subjectivities has been a largely unexplored theme. Studies of science and technology (including Latourian-inspired actor-network theory (ANT)) appear to have contributed more to the understanding of the connections between minute aspects of living and wider material influences than have mainstream sociological and cultural studies. Why has social theory found it difficult to see and to incorporate the material world of technology into the study of social change, particularly at the micro level of individuals, ordinary life and human interdependencies?

A variety of social theories, including the information society, post-Fordism, late capitalism, globalisation and postmodernity, all proclaim a radical transformation of contemporary society as a result of technological change. These theories encompass the automation of production, information and communication technologies, and a knowledge-based economy, affecting work, education, leisure, family relationships, cultural identities, sexualities, classifications of social living and other aspects. But I note, in agreement with Judy Wajcman (2002), that recent social theories addressing world transformations have a tendency to adopt a technological determinist stance. Technology is often placed as the most important cause of social change. This is the case with Manuel Castells (1997a), John Urry (2000) and Anthony Giddens (1991), for example. How do these approaches to technology and social change relate to the ways I have discussed technologies in this book?

Conceptions that social structure is technologically determined are strongly challenged in my adoption of ANT's claim that objects have 'agency'. I have sought to demonstrate that the material world lives *with* the human world, making one another. Social life is constituted in movement and technologies are constructed as a moving relational process achieved in daily social interactions (Law, 1999). Wajcman (2002: 305) notes that the action-oriented conception of ANT is reminiscent of ethnomethodology and chimes well with postmodern emphases on 'performativity' (Butler, 1993) derived from the work of Michel Foucault. How does this tension between the material and the social bear on my consideration of 'things' as resources, both theoretically and in my research practices?

Relations between the material and the social: 'things' as resources

My purpose has been to engage fully with the ordinary and work theoretically with these means. I have sought to consider objective relations with the material world and subjective experiences of the world (self, personal evaluations), employing the approach of ANT instrumentally, mostly from Latour: how to capture empirically the social world embedded in objects. I also employed the approach from Bourdieu both to capture differences in social hierarchies and to understand the patterns of differentiation on the basis of gender (for which Bourdieu is only marginally useful) and class. I explored technologies as part of personal and social relationships, using them in fieldwork as resources for my entry into conversations about individual practices and stories and as analytically crucial ingredients for forming the social.

Foucault (1988) refers to the technologies of the self, but his understanding of technology does not refer to the material. His concern is with social technologies and with the power relations embedded in them. But clearly an exploration of the power of objects and of the material in social and cultural life, while incorporating the understanding of technology as social, does not need necessarily to lose sight of the power of materiality. Surely, Foucaultian thinking incorporates the technical system into the social, where these are 'naturalised' in a process of exercising power by defining how objects work, what demands objects place upon users, how they define actors, how they construct relationships between the material and the human and between humans (Latour, 1992; Wajcman, 2002). While it is crucial to focus on the power that the social exerts via the material, Foucault's thinking in this regard is limited because of his failure to account for the effects on the social world of the social properties of intellectual producers (or the 'technologists' if we want to read this in ANT terms) and of their interrelations (see Lebaron, 2010 for an account of Bourdieu's critiques of Foucault).

I used both the ANT (Foucaultian and Latourian-inspired) frames of reference and the approaches inspired by Bourdieu to reflect on how technologies appear as 'relational resources'. In the application of these approaches to my concerns, I have taken an instrumental approach to ANT, using it to build on the idea of the connection between the world of technological objects and the inter-dynamics of individuals in their everyday dealings in their home life. The notion of 'technological scripts' is a central one which I adopted. I have also followed the prescription of 'following the actors' and 'tracing connections'. But, unlike Latour, I have operated within an ontological conception of the social as multilevel and multifaceted. While using descriptions, as he suggests, I have sought explanations, which Latour (2005) argues against. I followed Bourdieu's practice of articulation of ethnographic fieldwork with measures of 'objectification' in statistical data and other descriptions and analyses. I combined empirical observations with systematic comparisons, highlighting case studies which indicated in their singularity an accumulation of descriptions that revealed patterns or illustrated differences to confirm patterns. In this sense, and against Latour, my research reveals patterns, but it does not create patterns. While the individual stories have given a direction of social change, the stories are not of my making, though the analytic elaboration here presented is, of course, mine.

How can my reader be convinced of the strength of my interpretation? The concrete research operations I followed bear on the discussion of the cases, on my findings and their connections with other findings by other researchers. Bourdieu argues that the construction of the object of research analysis is made within the framework that creates the object. The method interrelates the empirical and the theoretical, and I further discuss below the connections I make about these.

To return to the connections between immediate occurrences and long-term changes, the evidence I discuss in Chapters 2, 4 and 5 regarding new imagined futures scripted in technological innovations show that these are strongly marked by existing social divisions and hierarchies. The stasis of domestic ideals has attempted to counter drives for change, reinstating social worlds which are hierarchically gendered and middle class. Wider social changes appear bounded by long-lasting relationalities, simultaneously reinstating some patterns and modifying others. However, relationalities are subject to constant change, although available resources condition different practices.

Resources in relationalities: concrete practices and empirical knowledge

Resources, some of which are technologies, construct social divisions because they are built within assumptions constituted in these divisions. That is not to say that technologies are not changed with use. Indeed they are, and the assumptions of their construction may be greatly transgressed. In Chapter 5, the contrasting cases of uses of technologies in the home show how similar machines designed to do the same tasks can be appropriated as resources for living in entirely different ways, with contradictory results. However, the reproductive technologies of artificial insemination used by Josie Barker (Chapter 8) resourced the formation of a particular multiethnic family conforming to her and her partner's choice of cultural insertion. Yet, the communication technologies used by Henry Gow (Chapter 8) had the disruptive effect of blowing up his long-guarded secret and disclosing his deceitful behaviour, thus failing as a resource for his social position as a respectful heterosexual married man, raising prospects of risk to his professional stand. Therefore, technologies are as complex and messy (Hughes, 2004) as the social worlds of which they are a part. As resources, technologies can operate as much as assets to enhance social position or to hinder one's position. Technologies are a part of personal and cultural relationalities.

My methods of investigation are, both materially and intellectually, part of the relationality here discussed. The engagement with people, their stories and doings, and my academic learning about these engagements, involved different kinds of intersections. Being attentive to these different sections and the connections between them helped me escape a compartmentalisation of knowledge. Hopefully my concerns with immediacy achieved a balance with the long term. Likewise, I hope to have broken the distinction between 'higher' and abstract understanding on the one hand, and 'lower' practice/activity, engaged with ordinarily, on the other.

My concern was to learn about social practices from fieldwork in family settings. To reflect again on some points I raised in the previous two sections: how can we describe these practices individually and yet consider patterns? How can we work out the ways in which individual practices bear on collective practices and resources?

I closely followed the biographies and the relationships between individuals. In doing this, the instructions of the psychosocial approach of Wendy Hollway and Tony Jefferson (2000) were productive. The sexual stories in my research (Chapter 8) are indicative of this approach, showing how the multidimensionality of the social was captured in the process of my fabrication of knowledge, and the deciphering of this process. The relationships that developed between research participants and me as a researcher were significant for the visibility of issues of intimacy of a sexual nature in the stories, for example, in accounts explicitly oriented to exploring relationships between people and technologies in daily domestic life. People liked telling stories, even when they were about discomforting events. However, the connections established do not solely come from the narratives. I remarked in Chapter 1 that the assumption that vision makes matter is a contentious one, since narratives that 'tell' do not necessarily 'show' (Bell, 1996; cf. Wittgenstein). Participants in the study told stories, but it was I, with my vision of linkages between certain aspects of social life, who showed the connections of the stories, or 'reassembled the social', to follow Latour's (2005) contention. Yet my assembly carries an ontological position which assumes a world where power marks social divisions and hierarchies.

To develop this further, it is important to emphasise that the concrete practices of individuals and of the researcher generating knowledge come into view in the context of ontological perspectives and their constructions of worlds. Thévenot's (2001) ideas are useful in this regard. He sees in people's accommodation to familiar environments the fine-tuning between people and between people and objects which, while bringing into view a habitus (situated and corporeal), involves a coordination of relationality. Personal creative dynamics are paramount here. Perhaps this view is not so distant from the one Bourdieu came to adopt in his later work, where the habitus appears as quite open, loose and flexible, unlike his examination in *Distinction* (Bourdieu, 1984). In *Pascalian Meditations*, Bourdieu (2000) argues that 'habitus may, in many cases, be confronted with conditions of actualization different from those in which they were produced ... Habitus change constantly in response to new experiences. Dispositions are subject to a kind of permanent revision, but one that is never radical' (pp. 160–1).

I do support a more flexible view of the habitus and I have found cases which were largely open to practices of creativity in the terms argued by Thévenot. Yet, constraints to creativity seem to appear on the basis of resources of cultural, social and emotional capitals. For instance, to revisit again Chapter 5: the practices followed by Rena Rock (with low cultural capital and depleted emotional resources) in her use and maintenance of household technologies and domestic routines would not fit with the contrasting case of Diane Churchill (with high cultural capital and emotionally resilient resources). It can be conceived that Rena could have had machines in working order and operated within efficient and orderly routines, but her subjective connections with her immediate world would have been affected. There is a connection between person and lifestyle which is not simply determined by class habitus. Emotional resources, creative adaptations and localised dynamics operate in this field. These are central to the exploration of the cases of Janet Seaman and Lynn Murray in Chapter 6. If relationships with machines, as relational resources, accrue differential advantages for the social positions of individuals and families depending on the positions they occupy in social space, the ability to make greater or lesser use of them varies according to personal resources and localised dynamics which are not exclusively or directly class-based.

Another important resource for social position concerns moral issues where ethical and emotional dimensions prevail. Drawing from Andrew Sayer (2010), I examined practices in which people acted not for advantage, but because they did think that certain courses of action were right or good in themselves. This contests Bourdieu's emphasis on competitive games and the *illusion* masking interests, and stresses the fact that ethical values and valuations are of enormous normative importance to people because they think that certain courses of action are valued. This also resonates with research by Luc Boltanski, Thévenot and Michèle Lamont on socially shared ways of classifying the social world. They see that the competing principles of justification drawn upon in different contexts (or what Lamont (2010) calls 'a type of cultural structure') emphasise fairness as fundamental in justifying social action. Within the relationality of families, fairness appears as a central concern, even when practices seem unequal in relation to other social fields.

The investigation discussed in this book indicates that the role of technology in home life is special. The role of culture in technology development and use is also special. Both technology and culture are relational resources embedded in the localised dynamics of the domestic which inform, and are informed by, wider socio-cultural and material practices. The resources of technology and culture are also produced by family relations: in practices, values, choices and compromises in everyday life, in past legacies and in the light of desired futures.

In terms of social change, the conundrum I note is that of practices, individual and social, using change to change, and of practices using change to stay the same. However, regardless of the intention, each practice changes the social: no permanence is possible. The diversity of stories contained in this book, including my own personal biographical engagements or those I make as a researcher, show us that we can recognise in individual lives the works of the culture, including the material technological culture, that forms the social world.

Appendix 1: Methodological Choices

This is a multidimensional study that includes two separate investigations on household technologies. One is based on archival research about selected technologies, while the other (most prominent in this book) is based on the ethnography of family life. I also examined secondary data and other studies to compare and place the research material in dialogue with wider and current concerns. I undertook the original investigation on household technologies in 1996, developing a frame for archival collection which I have continuously updated and enlarged to include newer technologies.[1] The earliest phase of the ethnography was done in 1998–9 when I interviewed and observed 16 households. Work on four other households was carried out in 2000 and on three others in 2001–2, the last one being included in 2004.[2] Various technological innovations have occurred since the late 1990s and inevitably a number of families engaged with the study in its earlier phases were not then having experiences which they might have had if they had been included later on. Secondary material is particularly relevant for this reason.

A lot of research work done in social sciences suffers from the need to respond to current concerns and this appears more pressing in areas where social change is perceived to be faster. While the core empirical material in this book was generated over a recent period, technological innovation and contemporary family life are two such areas. Because of this, and also because of the specific combination of materials produced for different purposes at different times, it seems that a reflection about re-using qualitative data is pertinent for my multi-use of research material. The re-use of data is a particularly energised methodology concern in the first decade of the new century in Britain (Moore, 2007) and I have engaged with this debate with an examination of the ethnographic part of the investigation that informs this book (Silva, 2007b). My position, which is shared with others (SRoL, 2007), is that despite the sense of urgency and the special status given to the issue, re-using data has been quite routine in social sciences. I stress that the processes of reusing data demand a critical reflection about what is included in the social world which is meant to be represented by the research and about what counts as evidence: the actors involved (here included objects) and their roles. I have a privileged position in relation to the use and reuse of the material in this book since I was involved in its generation. Space considerations preclude a detailed description of the research process, but in this appendix I describe some of the issues involved in the investigations included in this study.

Archival research and data sets[3]

For this project I utilised the methodology of applying semiotics to the process of data gathering and analysis for the study of technologies, as suggested by Madeleine Akrich and Bruno Latour (1992) (this is most explicit in the analyses in Chapters 4 and 5). Focusing on the needs of households and the market provisions, my leading questions were: who is the product for? What is expected from the machine and from those who operate it? Why? These questions centred on settings, scripts and programmes of actions, as identified through users' manuals, instructions and advice. The method enabled the analysis of the interplay of the intentions of the designer, manufacturer, market researcher and user. The technologies are also examined in relation to what they are for. Besides the concrete accomplishment of tasks, this examination also addresses the changes in morality implicit in particular lifestyles, including gendered labour, associated with specific activities. The mutual shaping of objects and practices involves a wide range of issues beyond the particular usage of the technologies and their roles in household relationships. The interplay of trends in the development of the technologies in both the industry and product markets is also emphasised.

The data sets produced initially for six household technologies – cooker, microwave oven, dishwasher, washing machine, fridge/freezer and disposable nappies – were later expanded to include computers, the Internet, mobile phones, televisions, video-recorders and DVD players. Two data sets were created: (1) a detailed analytic chronology of technical innovations in the selected technologies; and (2) a systematic survey of advertisements and reports on household technologies in women's magazines and feature pages of selected newspapers since the 1920s.

Ethnographic study

The study of families in a 'naturalistic' style is very much desired and rarely achieved. The original design and sampling of the study aimed to investigate home life in a 'natural' way by using video recordings. This was in 1997 and an ESRC award for a 'virtual ethnography' study offered an opportunity which was only partially successful.[4] This study provided some basis for observation of families in their own homes which was productively achieved when complemented by the more classic approach to ethnographic research (Silva, 2007b). Interviews and participant observation in the homes constitute the basic research material for the study of the families. Activities for five of the families were also explored by means of video recordings. The core of the observations and information from families in the 24 households derive from interviews carried out by me (17 families) and two research assistants (seven families) and the activities we engaged in with the participants and which we observed.

I will here describe some of the fieldwork choices concerning the sample and analysis.

All households include children of school age. The 16 households in the original sample for the 'virtual ethnography' were all white, married and heterosexual. These were recruited by a marketing research company and the families were paid a fee for access. These households were also video-recorded. I make very sparse use of these recordings, which did not yield significant material (for more information, see Silva, 2007b). The other eight households were recruited via personal contacts in a snowballing process and aimed to fill more specific criteria, as I intended to enlarge the sample to account for diverse styles of family living, including lesbian, lone parent and ethnic households. The resulting ethnic composition is of 18 white (including four Jewish), one Asian, one Afro-Caribbean and four mixed-race white and Afro-Caribbean. Four families have mixed nationality: two British and Canadian, one Jamaican and British and one German and British. Nearly one-third of the families conform to a non-conventional family type, defined as previously separated (cohabiting), lone parent and lesbian, with some of these characteristics appearing together, for example, as a lesbian lone parent. The sample is distributed between different areas of England with a mix of big and small towns including London (seven), locations in East Anglia (two), Lancashire (three), South Yorkshire (three) and North Yorkshire (six). The total interview sample includes 105 individuals: 25 women, 20 men and 60 children (27 girls and 33 boys). Since my purpose was to develop an intensive and detailed portrait of home life, I was able to include only a small number of families. The 45 adults were interviewed separately and the children were most often interviewed with other children in their family, as this was their preferred choice.

The fieldwork was done in the homes and consisted of extensive visits in the daytime and evenings, on weekdays and/or weekends and during holidays. Visits with each family lasted between six and 11 hours and many were revisited, up to three times, depending on their availability and enthusiasm. A structure of a set of questions was used with each interviewee and sets of equivalent data were completed on each case, focusing on particular topics. These evolved around a focused life story narrative. Although not imposing an ordered structure on the personal accounts, the themes I wanted to address were structured (see the last section of Chapter 1).

To analyse the data, I created a number of categories dealing with socio-demographics and relations with technologies. I also worked with each particular case study on its own, both at the level of the individual and of the household. The research was based on the ethical principles of getting consent and recognising the person as having their own feelings, capacities and wishes to reveal or conceal information. I renegotiated the process of eliciting information as information was being given. I have reflected, and carry on doing so, upon my biography, my relationship to the particular person, her home and her family, and how it felt like to be in this research relationship, as I have proceeded with the analysis and writing about the study.

In the analytic process, social class became a salient category for classification of the households. Here social class is interpreted as a package of assets which includes economic, social and cultural capital. These are unpacked in the analysis. The basic classification outlined in the profile of participants derives from the combined occupational class of the adults heading the household, or the one with the highest occupation, following the current British classificatory system (NS-SeC), grouped into three classes to reflect a stronger social and cultural conception of class than that focused only on the economic characteristics of occupations. The classification reflects the shape of the contemporary class structure (Bennett *et al.*, 2009), although they conceal important internal variations, as the discussion of individual cases in the book illustrate. Professional-executive households comprise professionals, managers in large establishments and large employers. The intermediate-class households include the lower managers, employers of small organisations and intermediate occupations. The working-class households include lower supervisors, technicians, those who account themselves as 'workers', semi-routine and routine occupations. I have also included 'poor' households, usually excluded from social class groupings, in this group. Table A1 shows the distribution of the sample. Table A2 shows special issues about ownership of technologies.

My intellectual journey from my emerging interest in technologies in the home until the writing of this book was not solely compounded of the research processes I outline here, but was greatly affected by changes in my personal life, including a change of job and the impact of new

Table A1 Households by social class and race/ethnicity

Class Race	Professional-executive	Intermediate class	Working class
White	H1 – **Green** Tracy and Gabriel + 2 children	H3 – **Addison** Brenda and Colin + 4 children	H2 – **Hughes** Katie and David + 2 children
	H5 – **Chambers** Rose and Ronald + 2 children ◘	H4 – **Lakin** Marion and Trevor + 2 children	H8 – **Seaman** Janet and Daniel + 3 children
	H9 – **Bird** Wendy and Scott + 2 children	H6 – **Goodman** Rosanne and Mike + 3 children	H15 – **Naylor** Jane and Uli + 4 children ◘
	H13 – **Webster** Chris and Phil + 4 children	H7 – **Mitchell** Nancy and Alfred + 1 child	
	H14 – **Hays-Field** Irene and Ian + 3 children	H10 – **Rock** Rena and John + 3 children	
	H16 – **Churchill** Diane and Marc + 3 children	H11 – **Gibson** Frances and Robert + 2 children	
	H17 – **Turner-Hill** Rebecca and Eleanor + 1 child •	H12 – **Wells** Lindsay and Ray + 6 children	
	H24 – **Lilly-Gow** Lucey and Henry + 1 child ◘		
Black		H19 – **MacDonald** Clare and Raj + 2 children	H20 – **Bartholomew** Richard + 1 child* (another child lives with mother)
Mixed race	H23 – **James-Cox** Jude and Anna + 1 child •	H22 – **Barker** Josie + 2 children • *	H18 – **Murray-Hall** Lynn + 4 children*
			H21 – **Al-Thompson** Lianne and Fred + 2 children ◘

• lesbian household.
* lone parent household.
◘ mixed nationality household.

Table A2   Special issues about ownership of technologies

Technology	Numbers	Length of use	To note
Computer	2 homes did not have any (H6, H8) 2 homes did not have it installed (H2 had an old one, H11 had only recently acquired one)	Over half had been acquired since 1995 Over a third had had a computer since before 1995	H1 and H16 worked from home. H17 and H24 partially worked from home. H9 male was a computer system designer
Internet	11 homes did not have an Internet connection	H1 had a connection since 1992 H12 and H16 had a connection since 1995	H1, H12 and H16 worked from home
Microwave oven	5 homes did not own one: H8, H17, H20, H22 and H24	Most homes were long-term users. Over half had owned one since 1995	For households that did not own one: H8 had had one for 8 years, H17 and H22 lesbian households, H20 on student grant, H24 had an Aga cooker
Dishwasher	7 homes did not own one: H12, H17, H18, H19, H20, H21, H22	Most who owned one had acquired it between 1989 and 1995	Homes with no dishwasher were all outside the 'norm' (H12 cohabiting)
Fridge/ freezer	All homes owned one, 3 homes had 2: H3, H9, H10	About half had acquired in last 5 years. Secondhand in H4 and H22	
Washing machine	All homes owned one	Most had been in possession for 2–10 years	
Tumble dryer	8 homes did not own one: H8, H11, H18, H19, H20, H21, H22, H23. H6 owned one but never used it. H14 owned a very small one and hardly used it. H2, H4 and H17 owned a washer/ dryer	Most in possession had had them for 2–10 years	Over half rarely used a tumble dryer

Table A2 (Continued)

Technology	Numbers	Length of use	To note
Television	Only H16 did not own one	Only H20 had one set. Others owned various sets with different 'ages': 6 sets: H3; 5 sets: H4, H12, H15	Always located in lounge + another room (kitchen or bedroom)
VCR/DVD player	All but H16 owned a VCR/DVD player. H16 had just acquired a CD-Video player in 1998	As a rule more recent ownership than TV. 4 sets: H4; 3 sets: H12, H13, H11. 5 homes had just one set	Location as for television set

academic themes, chiefly that on 'cultural capital and social exclusion', a project I developed with colleagues from 2002 to 2008 (Bennett *et al.*, 2009). I had begun the study being interested in historical changes and the particular way English families with children organised their daily lives. Yet, unexpected themes also emerged, in particular sexuality (Chapter 8), and the varied significance of individual and household resources for cultural capital and social class, developing some of my earlier concerns (Silva, 2006) and reinforcing findings from another study I co-authored (Bennett *et al.*, 2009).

When asked what I learned from all of this, I answer that as an academic I now know better that discipline boundaries offer little to the development of fields of enquiry on real dimensions of the social. I also know that a methodological approach employing a range of methods complementing one another offers the most productive means of addressing complex social phenomena (Silva, Warde and Wright, 2009). As a person living in Britain, I have experienced the welcome of strangers in their interest to share their stories and everyday habits, of showing how they do things and what sense they make of what they do. I learned much from the people in this study, which also helped me with my living in practical ways and in understanding my own practices, both as part of this culture and as different from it. As a person in the world I discovered that at many levels all the homes in this study felt like home. Many of these stories could be happening elsewhere, anywhere. A large proportion of the technologies appearing here could be simple objects used elsewhere, and it would still have a significant position in the relationships between individuals. Yet, there are also various particularly significant occurrences in these stories that are specific of British culture.

Appendix 2: Profile of Participants

For reference while reading, I provide profiles of participants. All names are pseudonyms and the brief demographic information and family situation and location aim to protect their identification. While the profiles offer specific information that might be of assistance in interpreting findings, considerable material from observation and interaction with participants is not included here. My aim was to tell a collective story focusing on individual situations. Income data refer to approximate *net* values.

H1 – Tracy (38) and Gabriel (41) **Green** were both architects. They lived in a large city in Yorkshire. She worked from home, officially part-time, but said work occupied her full time. He was director and partner at a practice in Lancashire. They had two sons, Oliver (8) and Ben (5). Both children were in private schools. They had two computers, two laptops, two television sets and were early Internet users. They owned all the other technologies. Income was £45,000 (woman's share = £15,000).

H2 – Katie (31) and David (30) **Hughes** lived in rural Yorkshire. She worked part-time in a supermarket carrying out shelf maintenance. He was a self-employed builder. They had two girls: Felicity (6) and Molly (4). They did not own a computer or a tumble dryer. Income was £14,000 (woman's share = £4,000).

H3 – Brenda (33) and Colin (35) **Addison** lived in a large Yorkshire town. She worked as a housewife and part-time fitness instructor for the City Council. He was a self-employed builder. They had three children in state school: Charlie (13), Daniela (11) and Eric (10) and a seven-month old baby Harry. They owned one computer but had no Internet connection. All other technologies were owned. Income was £37,000 (woman's share = £2,500).

H4 – Marion (43) and Trevor (42) **Lakin** lived in north London. They were Jewish. She worked as a cleaner, childminder and housewife. He was a bankrupted salesman. They had two boys, Keith (13) and Sean (8), in the local comprehensive. Sean also worked on a West End musical. They had all sorts of technologies including a computer and Internet access, but no tumble dryer. Income was £13,800 (woman's share = £2,800).

H5 – Rose (41) and Ronald (43) **Chambers** lived in north London. She was Canadian and worked as a school management officer for the Local

Authority. He was a classical music radio programme producer. Their children, Susie (10) and Steve (6), were in the local comprehensive school. They had one computer with Internet access, two television sets and all other technologies. Income was £59,000 (woman's share = £21,000).

H6 – Rosanne (37) and Mike (41) **Goodman** lived in North London. He was a sales manager in household textiles and she was a housewife who occasionally did some home-hairdressing. They had three children: Eliot (9), Tony (6) and Jonah (18 months). They were orthodox Jews and the children attended the Jewish school. They had no computer or Internet access, but owned two television sets and all the other technologies. Income was £25,000 (woman had no earnings).

H7 – Nancy (37) and Alfred (38) **Mitchell** lived in north London. She worked as a school lunchtime supervisor and a 'mystery' shopping researcher. He was a dental technician with his own practice. They were Jewish and had a son, Peter (8), who went to the local state school. They had one computer but no Internet access and owned all the other technologies. Income was £32,000 (woman's share = £4,000).

H8 – Janet (35) and Daniel (39) **Seaman** lived in a Yorkshire city. She worked as a tutor on cake decoration in community education and he was a boilermaker in a steel factory. Their three girls, Megan (10), Sophie (7) and Alex (4), attended the local state school. A childminder also looked after Alex. They did not own a computer, microwave oven or tumble dryer. Income was £24,000 (woman's share = £6,000).

H9 – Wendy (40) and Scott (40) **Bird** lived in East Anglia, in a small town. She was a housewife and private maths tutor who also volunteered with the local Red Cross. He was a computer systems designer for a large financial company. They had two children, Olivia (9) and Hugh (7), attending the local state school. They had all sorts of technology in the home, including two computers, one laptop and were early Internet users. Income was £35,000 (woman's share = £1,600).

H10 – Rena (44) and John (50) **Rock** lived in a small village in East Anglia. Rena worked as a housewife and occasional boat cleaner. John was an operator's manager for a large offshore company. They had three sons: Alan (17) studied at a private engineering college, while Geoff (14) and Patrick (12) were at the local state school. They owned all sorts of technologies and were early Internet and mobile phone users. Income was £40,000 (woman's share = £1,200).

H11 – Frances (40) and Robert (42) **Gibson** lived in rural Yorkshire. She worked as a bank official in a job-share position. He was sales manager for a telecommunications firm. Their children were George (9), who went to the local state school, and Emma (4), who went to the nursery in the same school as her brother. Computing technology had not made its way into the household by 1999. They had all other technologies but no tumble dryer. Income was £35,000 (woman's share = £9,000).

H12 – Lindsay (40) and Ray (45) **Wells** lived in a Lancashire town. She was a manager of an online retailer shop. He owned his own plumbing maintenance business. This was a second marriage for both and jointly they had six children who fully or partially lived in the household: Geoff (20), Cathy (14), Caroline (14), Vicky (11), Marcia (8) and Jack (4). They had a computer and Internet access, used mainly for business. All other technologies were owned. Income was £38,000 (woman's share = £10,000).

H13 – Chris (44) and Phil (46) **Webster** lived in a Lancashire town. She worked as a school crossing and playground supervisor and hairdresser from home. He was an actor, writer and director for theatre and television and was fairly well known for his work. They had four children in the local state schools: Greg (13), Georgia (11), Josh (8) and Joseph (6). The computer was lodged in the adults' bedroom. They had Internet access and all other technologies. Income was £25,000 (woman's share = £4,500).

H14 – Irene (39) and Ian (40) **Hays-Field** lived in a Lancashire town. She was a reference librarian in the public library. He was an engineer with a large engineering company. Their children, Katie (11) and twins James and Christopher (8), attended the local state school. They had two computers with modems and all sorts of technologies in the home. Income was £32,000 (woman's share = £9,000).

H15 – Jane (35) and Uli (35) **Naylor** lived in north London. She owned a small pre-school nursery and he was a safety and quality mechanical engineer for underground trains. He was of German descent. Their children, Will (13) and triplets Bob, Sonia and Bianca (7), attended the local state school. They did not have Internet access but owned all other technologies. Income was £27,000 (woman's share = £8,000).

H16 – Diane (43) and Marc (44) **Churchill** lived in a semi-detached house in north London. She was a lecturer in education and he was an academic Reader in information management. Their children, Greg (15), Hannah (11) and Alice (9), were in the local state schools. Kirk (41) was Diane's brother who lodged with them. They had four computers with Internet access. They had no television, though Kirk had one in his bedsit. All other technologies were owned. Income was £80,000 (woman's share = £33,000).

H17 – Rebecca **Turner** (46) and Eleanor **Hill** (51) lived in a small Yorkshire village. They were lesbians, cohabiting with Deborah (10), at state school, who was Rebecca's daughter from a previous lesbian relationship. Rebecca had a PhD and worked as an academic researcher. Eleanor worked as a university student advisor. They had three computers in the house, although they were late internet connectors. They did not own either a dishwasher or a tumble dryer. Income was £27,000 (half from each).

H18 – Lynn **Murray** (45) was white and had mixed race children (white Afro-Caribbean). She lived in a Yorkshire city and was lone mother of the **Hall** children: twins Gillian and Hayley (11), Chantal (8) and Sara (5). Lynn was unemployed and the family lived on benefits. They did not have a dishwasher, tumble dryer or Internet access. Income was £7,000 (from benefits).

H19 – Clare (33) and Raj (36) **MacDonald** lived in a large Yorkshire city. She was a secretary for a charity, working in a job-share position. He was a civil engineer. They were Asian (Indian descent). Their children, Katherine (10) and Nigel (6), were at state school. They owned a computer but had no Internet access. They did not own a dishwasher or a tumble dryer. Income was £30,000 (woman's share = £10,000).

H20 – Richard **Bartholomew** lived in a Yorkshire city. He was a black Afro-Caribbean lone father of Hazel (8), who lived with her mother, and Thomas (15), who attended the local comprehensive and lived with him. He was a postgraduate student. He owned a computer but had no Internet access. Income was £8,000 (from a student grant).

H21 – Lianne **Al** (40) and Fred **Thompson** (38) lived in a Yorkshire city. She worked as a part-time jewellery returns clerk for a mail order company. He was on incapacity benefit, doing an Open University degree. Lianne was white and British and Fred was Afro-Caribbean from Jamaica. Their children, Adam (9) and Edward (3), attended the Catholic school and nursery near their home. They had a computer and Internet access, but no dishwasher or tumble dryer. Income was £11,400 (about half from each, Fred's share coming from benefits).

H22 – Josie **Barker** (42) lived in south London with her two children, Michael (11) and Cassie (4). He went to the local comprehensive school and she was looked after by a combination of childminders and nursery. Josie worked as a trade union officer for lesbian and gay issues – she was a lesbian, white, and her children were mixed race (white Afro-Caribbean). She owned a computer with Internet access, but no microwave oven, dishwasher or tumble dryer. Income was £20,000.

H23 – Jude **James** (37) and Anna **Cox** (38) lived in a Yorkshire town. Jude was a full-time teacher (head of her subject) and Anna was a literacy consultant to the Local Authority. They cohabited with Anna's son from a previous relationship, Alex (9), who went to the local state school. They had an 'Internet room' with two computers and a third computer in the child's bedroom. They did not own a tumble dryer. The women were white and the child was mixed race (white Afro-Caribbean). Income was £46,000 (about half from each).

H24 – Lucey **Lilly** (50) and Henry **Gow** (53) lived in a large Lancashire city. They were white, married and had a son, Stuart (14). She was a solicitor and he was an orthopaedic consultant. They had one computer and three laptops with Internet access and owned all other technologies. She was Canadian. Income was £82,000 (woman's share = £34,000).

Notes

Chapter 1

1. I use the term 'innovator' to designate the unspecified assembly of designers, manufacturers and marketing professionals implicated in technological innovation processes.
2. I use the expression 'talk' to indicate ways of communicating with and through. I am interested in the relationality between objects and individuals and also about the researcher relationship in the investigation of these relations.
3. The environmental and animal worlds are, of course, players in this field, as usefully reflected on in Giffney and Hird (2008).

Chapter 2

1. I chose *net* income values because most women's earnings were not taxable and the idea of 'income brought home' was easier to talk about and compare.
2. In another study I carried out with colleagues, we found that wives of 'elite' men tended to display more traditional gender roles, as did the husbands (Bennett *et al.*, 2009: Chapter 12).
3. Of course, in saying this I do not assume that the undoing of traditional arrangements of gender implies an end to male domination. This may simply be reworked.

Chapter 3

1. I draw here from arguments I developed in an earlier paper: Silva, 2002a.
2. Overall, time spent cooking and cleaning from the 1960s to the mid-1990s in 20 North American and European countries decreased for women by just under one hour per day, while men's time increased by 20 minutes daily (Sullivan, 2000; cf. Gershuny, 2000).
3. A similar trend was observed in Australia by Bittman, Matheson and Meagher (1999).
4. Household H24 included in the study in 2004 constitutes a particular case for which routines in the home were not explored. There is no full interview for the man in household H2. In two households, partners were of the same sex.
5. This replicates some points from an earlier analysis of a smaller sample: Silva, 2002b.

Chapter 4

1. For a more detailed exploration of these issues, see Silva, 2000a.
2. Most references are housed at the Blanche Leigh Special Cookery Collection, Brotherton Library at the University of Leeds. See also the University of

Cambridge Central Library, the British Newspaper Library and the National Magazine Company Library.

3. In electric cookers, probably the first electric thermostat control, the Revostat, was introduced in 1935 by the Revo Electric Company Ltd.

4. In the censuses, in 1951 private domestic service amounted to only 15 per cent of the 1931 level.

5. Corley (1966: 111) notes, for instance, that pre-War electric cookers even retained the familiar but cumbersome knobs and bars inherited from the gas stove, while improved 'snap' handles were not introduced until after the War.

6. Cowan's (1983) brilliant analysis of four centuries of household technological development includes a chapter on 'The Roads Not Taken', with references to some missed alternatives of this kind in the American market.

7. On the decline of home cooking in the USA, see Pillsbury, 1998: 187–97.

8. An Aga is a cast-iron heat-storage cooker that runs on gas, electricity, oil or coal (older models), allowing great versatility in cooking and a permanently warm kitchen, suited to cold weather countries. It costs an average of £3,000 for a basic four door model.

9. Marks & Spencer (M&S) is a high street shop chain which offers upmarket cooked and pre-prepared meals. It has become synonymous with taste and ease in eating at home.

10. The trends in my study broadly converge with those found by Frances Short (2006) in her interesting study of cooking as one of the 'kitchen secrets'.

Chapter 5

1. Notable engagements in the field of domestic cleaning are found in Ackerley (1994) and Martens and Scott (2005).

2. See details on the historical developments of technologies in Silva, 1997a.

Chapter 6

1. I discuss this case with a particular focus on gender and consumption in Silva, 2000b and 2007.

2. Habitat is a middle-class design furniture and home accessories shop.

3. I give an account of this case in a previous paper (Silva, 2000b). The original narrative is from Emma Heron.

Chapter 7

1. In the construction of my vignettes I closely followed Janet Finch's (1987) advice.

2. The cases did not allow for detailed discussion of other identity issues which might have affected these situations.

3. The issue of relevance brings about the example of Latour's (1996) rich study of the failed development of the electric car system in 1980s Paris. Perhaps the difficulties of some individuals to operate VCRs led to designs for an easier operation of DVD players.

Chapter 8

1. My use of the concepts of 'noisy' and 'muted' sexuality aims to capture the different registers in the politics of pleasure between extraordinariness and ordinariness implied in practices that are in the first case visible (or that achieve visibility when a 'secret' practice is disclosed) against the normalisation and invisibility prevailing in the latter.
2. The PAS was the central London clinic carrying treatments of the British Pregnancy Advisory Service (BPAS), which stored donor sperm before the Human Fertilisation and Embryology Act came into force in 1993.
3. Josie's net income was about £20,000 and she did not own her London flat. Rena's household net income was about £40,000 and their East Anglian house was fully owned.
4. My initial contact with this family is from 2002, although their formal inclusion in the study dates from 2004.
5. 'Cookies' are a mechanism by which website operators can place a unique identifier on each user's machine. The site can then compile a history of each user's activities across browser sessions.
6. Numerous cases of discoveries of Internet practices involving sex have been reported. Examples include: (1) Shock at husband's police arrest for downloading child porn (*Observer, First Person*, 5 August 2007). (2) Husbands looking at porn sites daily for years on end (*The Guardian, G2*, 26 May 2008). (3) Family negotiations to protect children from pornography use by fathers (*The Guardian, Family*, 27 August 2008). (4) Real life divorce following husband being caught on the *Second Life* site, where adults can fantasise alternative selves, bodies and lives, shagging a computerised beauty who he planned to marry in real life (*The Guardian, G2*, various dates, November 2008).

Appendix 1

1. I benefitted from CRESC (ESRC Centre for Research on Socio-cultural Change) funding for research updating and Hannah Knox assisted me with this work in 2006–7.
2. Funding for the expanded sample was provided by the School of Sociology and Social Policy, University of Leeds and the National Everyday Cultures Programme in Sociology at the Open University. Emma Heron assisted with the work in 2000 and Pippa Stevens in 2001–2.
3. This study was funded by the ESRC (award no. R000231700). For a fuller account of the material reviewed, see Silva, 1997b.
4. The study was funded by the ESRC under the 'Virtual Society?' Research Programme (award no. L132251048). Details about this experiment can be found in Silva, 1999a and 2007b.

Bibliography

Ackerley, L. (1994) 'Consumer Awareness of Food Hygiene and Food Poisoning', *Environmental Health*, March, 69–74.

Adam, B. (1995) *Timewatch. The Social Analysis of Time*. Cambridge: Polity.

Adkins, L. (2002) *Revisions: Gender and Sexuality in Late Modernity*. Buckingham: Open University Press.

Adkins, L. (2005) 'The New Economy, Property and Personhood', *Theory, Culture & Society* 22(1), 111–30.

Adkins, L. (2009) 'Feminism after Measure', *Feminist Theory* 10(3), 323–39.

Aglietta, M. (1979) *A Theory of Capitalist Regulation*. London: New Left Books.

AHS (2008) *American Housing Survey for the United States*: 2007 ix – U.S. Department of Housing and Urban Development and U.S. Census Bureau. Table 1A-4. 'Selected Equipment and Plumbing_All Housing Units'. Page 5. http://www.census.gov/prod/2008pubs/h150-07.pdf, accessed 27 March 2009.

Akrich, M. (1987) 'Comment décrire les objects techniques', *Technique et Culture*, Jan–Jun, 49–63.

Akrich, M. (1992) 'The De-Scription of Technical Objects', in W. Bijker and J. Law (eds) *Shaping Technology/Building Society*. Cambridge, Mass: MIT Press.

Akrich, M. and Latour, B. (1992) 'A Summary of a Convenient Vocabulary for the Semiotics of Human and Non-human Assemblies', in W. Bijker and J. Law (eds) *Shaping Technology/Building Society*. Cambridge, Mass: MIT Press.

Appadurai, A. (1986) 'Introduction: Commodities and the Politics of Value', in A. Appadurai (ed.) *The Social Life of Things*. Cambridge University Press.

Attfield, J. (2000) *Wild Things. The Material Culture of Everyday Life*. Oxford: Berg.

Attwood, F. (2002) 'Reading Porn: The Paradigm Shift in Pornographic Research', *Sexualities* 5(1), 91–105.

Attwood, F. (2009) '"Deepthroatfucker" and "Discerning Adonis". Men and Cybersex', *Sexualities* 12(3), 279–94.

Baden-Fuller, C.W.F. and Stopford, J.M. (1992) *Rejuvenating the Mature Business. The Competitive Challenge*. London and New York: Routledge.

Barret, M. (1992) 'Words and Things: Materialism and Method in Contemporary Feminist Analysis' in M. Barret and A. Phillips (eds) *Destabilizing Theory. Contemporary Feminist Debates*. Cambridge: Polity.

Bauman, Z. (1995) *Life in Fragments*. Oxford: Blackwell.

Beck, U. (1992) *Risk Society: Towards a New Modernity*. London: Sage.

Beck-Gernsheim, E. (1998) *Reinventing the Family. In Search of New Lifestyles*. Cambridge: Polity.

Bell, V. (1996) 'Show and Tell: Passing and Narrative in Toni Morrison's *Jazz*', *Social Identities* 2(2), 221–36.

Bennett, T. and Frow, J. (2008) 'Introduction: Vocabularies of Culture', in T. Bennett and J. Frow (eds) *The Sage Handbook of Cultural Analysis*. London: Sage.

Bennett, T. and Silva, E.B. (2004) 'Everyday Life in Contemporary Culture', in E.B. Silva and T. Bennett (eds) *Contemporary Culture and Everyday Life*. Durham: Sociologypress.

Bennett, T., Savage, M., Silva, E., Warde, A., Gayo-Cal, M. and Wright, D. (2009) *Culture, Class, Distinction*. London: Routledge.

Berg, A-J. (1996) *Digital Feminism*. Centre for Technology and Society, Norwegian University of Science and Technology, Report no. 28.

Birdwell-Pheasant, D. and Lawrence-Zúñiga, D. (1999) 'Introduction: Houses and Families in Europe', in D. Birdwell-Pheasant and D. Lawrence-Zúñiga (eds) *House Life. Space, Place and Family in Europe*. Oxford: Berg, pp. 1–35.

Bittman, M., Matheson, G. and Meagher, G. (1999) 'The Changing Boundary Between Home and Market: Australian Trends in Outsourcing Domestic Labour', *Work, Employment and Society* 13(2), 249–73.

Bose, C. (1979) 'Technology and Changes in the Division of Labour in the American Home', *Women's Studies International Quarterly* 2, 295–304.

Boltanski, L. and Thévenot, L. (2006) *On Justification. Economies of Worth* (trans. Catherine Porter). Princeton University Press.

Bourdieu, P. (1977) *Outline of a Theory of Practice*. Cambridge: Polity.

Bourdieu, P. (1984 [1979]) *Distinction. A Social Critique of the Judgement of Taste*. London: Routledge.

Bourdieu, P. (1988) *Homo Academicus*. Cambridge: Polity.

Bourdieu, P. (1992a) *The Logic of Practice*. Cambridge: Polity (orig. *Le Sens Pratique*, 1980).

Bourdieu, P. (1992b) 'The Kabile House of the World Reversed', Appendix in *The Logic of Practice*. Cambridge: Polity.

Bourdieu, P. (1993 [1984]) *Sociology in Question*. London: Sage.

Bourdieu, P. (1996 [1992]) *The Rules of Art: Genesis and Structure of the Literary Field*. Cambridge: Polity.

Bourdieu, P. (1998) *Practical Reason*, Cambridge: Polity.

Bourdieu, P. (1998b) *On Television*. New York: The New Press.

Bourdieu, P. (2000 [1997]) *Pascalian Meditations*, Cambridge: Polity.

Bourdieu, P. (2004) *Science of Science and Reflexivity*. Cambridge: Polity.

Bourdieu, P. (2005) *The Social Structures of the Economy*. Cambridge: Polity.

Bourdieu, P. and Wacquant, L. (1992) *An Invitation to Reflexive Sociology*. Chicago University Press.

Bourdieu, P. *et al.* (1999 [1993]) *The Weight of the World. Social Suffering in Contemporary Society*. Stanford University Press.

Brannen, J., Moss, P. and Mooney, A. (2004) *Working and Caring over the Twentieth Century*. Basingstoke: Palgrave Macmillan.

Braverman, H. (1974) *Labour and Monopoly Capital. The Degradation of Work in the Twentieth Century*. New York and London: Monthly Review Press.

Butler, J. (1990) *Gender Trouble*. London: Routledge.

Butler, J. (1993) *Bodies that Matter. On the Discursive Limits of Sex*. New York: Routledge.

Callon, M. (1986a) 'Some Elements of a Sociology of Translation: Domestication of the Scallops and the Fishermen of St Brieuc Bay', in J. Law (ed.) *Power, Action and Belief: A New Sociology of Knowledge?* London: Routledge.

Callon, M. (1986b) 'The Sociology of an Actor-Network: The Case of the Electric Vehicle', in M. Callon, J. Law and A. Rip (eds) *Mapping the Dynamics of Science and Technology*. Basingstoke: Macmillan.

Callon, M. (1989) 'Society in the Making: The Study of Technology as a Tool for Sociological Analysis', in W. Bijker, T. Hughes and T. Pinch (eds) *The Social Construction of Technological Systems*. Cambridge, Mass: MIT Press.

Callon, M. and Latour, B. (1981) 'Unscrewing the Big Leviathan: How Actors Macro-structure Reality and How Sociologists Help Them to Do So', in K. Knorr-Cetina and A.V. Cicourel (eds) *Advances in Social Theory and Methodology: Towards an Integration of Micro- and Macro-sociologies*. London: Routledge & Kegan Paul.

Casey, E. and Martens, L. (2007) 'Introduction', in E. Casey and L. Martens (eds) *Gender and Consumption: Domestic Cultures and the Commercialization of Everyday Life*. Aldershot: Ashgate, pp. 1–11.

Castells, M. (1997a) *The Information Age: Economy, Society and Culture. Network Society*. Volume I. Oxford: Blackwell.

Castells, M. (1997b) *The Information Age: Economy, Society and Culture. The Power of Identity*. Volume II. Oxford: Blackwell.

Chabaud-Rychter, D. (1994) 'Women Users in the Design Process of a Food Robot: Innovation in a French Domestic Appliance Company', in C. Cockburn and R. Fürst-Dilic (eds) *Bringing Technology Home: Gender and Technology in a Changing Europe*. Buckingham: Open University Press.

Chabaud-Rychter, D. (1995) 'The Configuration of Domestic Practices in the Designing of Household Appliances', in K. Grint and R. Gill (eds) *The Gender-Technology Relation*. London: Taylor & Francis.

Chabaud-Rychter, D. and Gardey, D. (eds) (2002) *L'engendrement des choses. Des homes, des femmes et des techniques*. Paris: Editions des Archives Contemporaines.

Charles, N. and Kerr, M. (1988) *Women, Food and Families*. Manchester University Press.

Cheng, S-L, Olsen, W, Southerton, D. and Warde, A. (2007) 'The Changing Practice of Eating: Evidence from UK Time Diaries, 1975 and 2000', *British Journal of Sociology* 58(1), 39–61.

Cockburn, C. (1985) 'The Material of Male Power', in D. Mackenzie and J. Wajcman (eds) *The Social Shaping of Technology*. Milton Keynes: Open University Press.

Cockburn, C. (1992) 'The Circuit of Technology: Gender, Identity and Power', in R. Silverstone and E. Hirsch (eds) *Consuming Technologies*. London: Routledge.

Cockburn, C. and Ormrod, S. (1993) *Gender and Technology in the Making*. London: Sage.

Cockburn, C. and Fürst-Dilic, R. (eds) (1994) *Bringing Technology Home. Gender and Technology in a Changing Europe*. Buckingham: Open University Press.

Connell, R.W. (2002) *Gender*. Cambridge: Polity.

Consumer Reports – various years – USA.

Corley, T.A.B. (1966) *Domestic Electrical Appliances*. London: Jonathan Cape.

Cowan, R.S. (1974) 'A Case Study of Technology and Social Change: The Washing Machine and the Working Wife', in M. Hartman and L. Banner (eds) *Clio's Consciousness Raised: New Perspectives on the History of Women*. New York: Harper and Row.

Cowan, R.S. (1983) *More Work for Mother: The Ironies of Household Technology from the Open Hearth to the Microwave*. New York: Basic Books.

Cowan, R.S. (1985) 'How the Refrigerator got its Hum', in D. Mackenzie and J. Wajcman (eds) *The Social Shaping of Technology*. Milton Keynes: Open University Press.

Dant, T. (1999) *Material Culture in the Modern World*. Buckingham: Open University Press.

Das, V. (2007) *Life and Words. Violence and the Descent into the Ordinary*. Berkeley: University of California Press.

De Certeau, M. (1984) *The Practice of Everyday Life*. Berkeley: University of California Press.

DeVault, M.L. (1990) 'Talking and Listening from Women's Standpoint: Feminist Strategies for Interviewing and Analysis', *Social Problems* 37, 96–116.

DeVault, M.L. (1991) *Feeding the Family. The Social Organization of Caring as Gendered Work*. London and Chicago: University of Chicago Press.

Douglas, M. (1966) *Purity and Danger: An Analysis of the Concepts of Pollution and Taboo*. London: Routledge & Kegan Paul.

Douglas, M. (1986) *Risk Acceptability According to the Social Sciences*. London: Routledge.

Dovey, K. (2002) 'The Silent Complicity of Architecture', in J. Hillier and E. Rooksby (eds) *Habitus: A Sense of Place*. Aldershot: Ashgate.

Easlea, B. (1983) *Fathering the Unthinkable: Masculinity, Scientists and the Nuclear Arms Race*. London: Pluto Press.

Edwards, J. (1999 [1993]) 'Explicit Connections: Ethnographic Enquiry in Northwest England', in J. Edwards, S. Franklin, E. Hirsch, F. Price and M. Strathern, *Technologies of Procreation. Kinship in the Age of Assisted Conception*. London: Routledge, pp. 60–85.

Edwards, J., Franklin, S., Hirsch, E., Price, F. and Strathern, M. (1999 [1993]) *Technologies of Procreation. Kinship in the Age of Assisted Conception*. London: Routledge.

EIA (2004) *The Effect of Income on Appliances in U.S. Households*, Energy Information Administration, http://www.eia.doe.gov/emeu/recs/appliances/appliances. html, accessed 26 March 2009.

Environment Agency (2008) *An Updated Lifecycle Assessment Study for Disposable and Reusable Nappies*. Science Report – SC010018/SR2, http://randd.defra.gov.uk/ Document.aspx?Document=WR0705_7589_FRP.pdf, accessed 3 April 2009.

Family Expenditure Survey (2001) ONS, London: The Stationery Office.

Family Spending (1997) A Report on the 1996–97 Family Expenditure Survey. ONS, London: The Stationery Office.

Faulkner, W. (2002) 'The Power *and* the Pleasure? A Research Agenda for "Making Gender Stick" to Engineers', *Science, Technology and Human Values* 25(1), 87–119.

Featherstone, M. (1991) *Consumer Culture and Postmodernism*. London: Sage.

Feenberg, A. (1999) *Questioning Technology*. New York: Routledge.

Finch, J. (1987) 'Research Note: The Vignette Technique in Survey Research', *Sociology* 21(1), 105–14.

Finch, J. (1989) *Family Obligations and Social Change*. Cambridge: Polity.

Finch J. and Mason J. (1993) *Negotiating Family Responsibilities*. London: Routledge.

Folbre, N. (1994) *Who Pays for the Kids? Gender and the Structures of Constraint*. London: Routledge.

Forty, S. (1975) *Objects of Desire*. London: Cameron Books.

Foucault, M. (1978 [1976]) *The History of Sexuality. Vol 1: An Introduction* (trans. Rob Harley). New York: Pantheon.

Foucault, M. (1988) 'Technologies of the Self', in L.H. Martin, H. Gutman and R.H. Hutton (eds) *Technologies of the Self*. London: Tavistock.

Franklin, S. (1999 [1993]) 'Making Representations: The Parliamentary Debate on the Human Fertilisation and Embryology Act', in J. Edwards, S. Franklin, E. Hirsch, F. Price and M. Strathern, *Technologies of Procreation. Kinship in the Age of Assisted Conception.* London: Routledge, pp. 127–65.

Fraser, M. (1999) 'Classing Queer. Politics in Competition', *Theory, Culture and Society* 16(2), 107–31.

Frederick, C. (1919) *Household Engineering: Scientific Management in the Home.* American School of Home Economics.

Friends of the Earth (2009) 'Dishwasher versus Hand Washing', http://www.foe.co.uk/living/articles/dishwasher_handwash_5915.html, accessed 4 April 2009.

Gabb, J. (2008) *Researching Intimacy in Families.* Basingstoke: Palgrave Macmillan.

Gardiner, J. (1997) *Gender, Care and Economics.* Basingstoke: Macmillan.

Gerodetti, N. and Mottier, V. (2009) 'Feminism(s) and the Politics of Reproduction: Introduction to Special Issue on "Feminist Politics of Reproduction"', *Feminist Theory* 10(2), 147–52.

Gershuny, J. (1983) *Social Innovation and the Division of Labour.* Oxford University Press

Gershuny, J. (2000) *Changing Times. Work and Leisure in Postindustrial Society.* Oxford University Press.

Gershuny, J., Godwin, M. and Jones, S. (1994) 'The Domestic Labour Revolution: A Process of Lagged Adaptation', in M. Anderson, F. Bechhofer and J. Gershuny (eds) *The Social and Political Economy of the Household.* Oxford University Press.

Gershuny, J., Lader, D. and Short, S. (2006) *The Time Use Survey, 2005: How We Spend Our Time. Time Use Results for 2005 Where Appropriate Compared with the UK 2000 Time Use Survey.* ONS, http://www.statistics.gov.uk/articles/nojournal/time_use_2005.pdf, accessed 24 February 2009.

GHS (1995) *General Househod Survey 1993.* Office of Population Censuses and Surveys, London: HMSO.

GHS (1998) *General Household Survey.* London: HMSO.

Giddens, A. (1991) *Modernity and Self Identity. Self and Society in the Late Modern Age.* Cambridge: Polity.

Giddens, A. (1992) *The Transformation of Intimacy: Sexuality, Love and Eroticism in Modern Societies.* Cambridge: Polity.

Giedion, S. (1948) *Mechanization Takes Command.* Oxford University Press.

Giffney, N. and Hird, M. (eds) (2008) *Queering the Non/Human.* Aldershot: Ashgate.

Gilligan, C. (1982) *In a Different Voice.* Cambridge, Mass: Harvard University Press.

Goffman, E. (1959) *The Presentation of the Self in Everyday Life.* New York: Doubleday.

Gomez, M.C.A. (1994) 'Bodies, Machines and Male Power', in C. Cockburn and R. Fürst-Dilic (eds) *Bringing Technology Home.* Buckingham: Open University Press.

Good Housekeeping (magazine) – various years and numbers.

Graham, E.L. (2002) *Representations of the Post/Human: Monsters, Aliens and Others in Popular Culture.* Manchester University Press.

Green, E. (2001) 'Technology, Leisure and Everyday Practices' in E. Green and A. Adam (eds) *Virtual Gender. Technology, Consumption and Identity.* London: Routledge.

Griffiths, M. (1995) *Feminisms and the Self.* London: Routledge.

Gronow, J. and Warde, A. (2001) *Ordinary Consumption.* Reading: Harwood.

Hacker, S. (1989) *Pleasure, Power and Technology.* Boston: Unwin Hyman.

Haimes, E. (1998) 'Changing Representations of the Child', in K. Daniels and E. Haimes (eds) *International Social Science Perspectives on Donor Insemination.* Cambridge University Press.

Hall, C. (1980) 'The History of the Housewife', in E. Mallos (ed.) *The Politics of Housework.* London: Allison and Busby.

Haraway, D. (1988) 'Situated Knowledges: The Science Question in Feminism and the Privilege of Partial Perspectives', *Feminist Studies,* 575–99.

Haraway, D. (1991) *Simians, Cyborgs and Nature.* London: Free Association Books.

Harre, R. (2002) 'Material Objects in Social Worlds', *Theory, Culture and Society* 19(5/6), 23–33.

Hochschild, A.R. (1983) *The Managed Heart, The Commercialization of Human Feeling.* Berkeley: University of California Press.

Hochschild, A.R. (1997) *The Time Bind,* New York: Metropolitan Books.

Hollway, W. (2006) *The Capacity to Care. Gender and Ethical Subjectivity.* London: Routledge.

Hollway, W. and Jefferson, T. (2000) *Doing Qualitative Research Differently.* London: Sage.

Hollows, J. (2008) *Domestic Cultures.* Maidenhead: McGraw-Hill and Open University Press.

Housewife (magazine) – various years and numbers.

Hoy, S. (1995) *Chasing Dirt. The American Pursuit of Cleanliness.* Oxford University Press.

Hughes, T. (1983) *Networks of Power: Electrification in Western Society, 1880–1930.* Baltimore: Johns Hopkins University Press.

Hughes, T. (2004) *Human Built World: How to Think about Technology and Culture.* University of Chicago Press.

Jackson, S. (2008) 'Ordinary Sex', *Sexualities* 11(1/2), 33–7.

Jackson, S. and Moore, S. (eds) (1995) *The Politics of Domestic Consumption.* Hemel Hempstead: Prentice Hall/Harvester Wheatsheaf.

Jackson, S. and Scott, S. (2004) 'Sexual Antinomies in Late Modernity', *Sexualities* 7(2), 233–48.

Jamieson, L. (1998) *Intimacy. Personal Relationships in Modern Societies.* Cambridge: Polity.

Joerges, B. (1999) 'Do Politics have Artefacts?' *Social Studies of Science* 28(3), 411–32.

Juffer, J. (1998) *At Home with Pornography: Women, Sexuality and Everyday Life.* New York University Press.

Juffer, J. (2004) 'Sexual Technologies/Domestic Technologies: Pornography as an Everyday Matter', in E.B. Silva and T. Bennett (eds) *Contemporary Culture and Everyday Life.* Durham: Sociologypress.

Kaufmann, J-C. (1998 [1992]) *Dirty Linen: Couples and their Laundry.* London: Middlesex University Press.

Keller, E.F. (2000) *The Century of the Gene.* Cambridge, Mass: Harvard University Press.

Key Notes (1996) *Household Appliances (White Goods).* Market Report (edited by R. Caines). Eleventh edition. London: Key Note Ltd.

Kimmel, M.S. (1994) 'Masculinity as Homophobia: Fear, Shame and Silence in the Construction of Gender Identity', in H. Brod and M. Kaufman (eds) *Theorizing Masculinities.* Thousand Oaks: Sage, pp. 119–41.

Lahire, B. (2004) *La culture des individus. Dissonances culturelles et distinction de soi*. Paris: Éditions La Découverte.

Lally, E. (2002) *At Home with Computers*. Oxford: Berg.

Lamont, M. (2000) *The Dignity of Working Men. Morality and the Boundaries of Race, Class and Immigration*. Cambridge, Mass: Harvard University Press.

Lamont, M. (2010) 'Looking Back at Bourdieu', in E. Silva and A. Warde (eds) *Cultural Analysis and Bourdieu's Legacy: Settling Accounts and Developing Alternatives*. London: Routledge.

Lang, T. and Caraher, M. (2001). 'Is There a Culinary Skills Transition?: Data and Debate from the UK about Changes in Cooking Culture', *Journal of the Home Economics Institute of Australia* 8(2), 2–14.

Latour, B. (1986) 'The Powers of Association', in J. Law (ed.) *Power, Action and Belief: A New Sociology of Knowledge?* London: Routledge.

Latour, B. (1987) *Science in Action. How to Follow Scientists and Engineers through Society*. Cambridge, Mass: Harvard University Press.

Latour, B. (1988a [1984]) *Pasteurization of France*. Cambridge, Mass: Harvard University Press.

Latour, B. (1988b) 'Mixing Humans and Nonhumans Together: The Sociology of a Door Closer', *Social Problems* 35, 298–310.

Latour, B. (1992) 'Where are the Missing Masses? The Sociology of a Few Mundane Artifacts', in W. Bijker and J. Law (eds) *Shaping Technology/Building Society*. Cambridge, Mass: MIT Press, pp. 225–58.

Latour, B. (1996 [1992]) *Aramis or the Love of Technology*. Cambridge, Mass: Harvard University Press.

Latour, B. (1999) *Pandora's Hope. Essays on the Reality of Science Studies*. Cambridge, Mass: Harvard University Press.

Latour, B. (2005) *Reassembling the Social. An Introduction to Actor-Network-Theory*. Oxford University Press.

Latour, B. and Woolgar, S. (1979) *Laboratory Life. The Social Construction of Scientific Facts*. London: Sage.

Laumann, E.O., Gagnon, J.H., Michael, R.T. and Michaels, S. (1994) *The Social Organization of Sexuality: Sexual Practices in the United States*. University of Chicago Press.

Law, J. (1987) 'Technology and Heterogeneous Engineering', in W. Bijker, T. Hughes and T. Pinch (eds) *The Social Construction of Technological Systems*. Cambridge, Mass: MIT Press.

Law, J. (1999) 'After ANT: Complexity, Naming and Topology' in J. Law and J. Hassard (eds) *Actor Network Theory and After*. Oxford: Blackwell and the Sociological Review.

Law, J. (2004) *After Method: Mess in Social Science Research*. London: Routledge.

Law, J. and Urry, J. (2004) 'Enacting the Social', *Economy and Society* 33(3), 390–410.

Layder, D. (1994) *Understanding Social Theory*. London: Sage.

Layder, D. (1996) 'Contemporary sociological theory', *Sociology* 30(3), 601–8.

Layder, D. (2004) *Emotion in Social Life. The Lost Heart of Society*. London: Sage.

Lebaron, F. (2010) 'Bourdieu in a Multidimensional Perspective', in E. Silva and A. Warde (eds) *Cultural Analysis and Bourdieu's Legacy: Settling Accounts and Developing Alternatives*. London: Routledge.

Lebergott, S. (1993) *Pursuing Happiness. American Consumers in the Twentieth Century*. Princeton University Press.

Lefaucher, N. (1995) 'De la stabilité à la mobilité conjugale', *Politis*, La Revue 8, Montreuil, France, pp. 19–23.

Lerman, N., Olderziel, R. and Mohun, A. (eds) (2002) *Gender and Technology. A Reader.* Baltimore: Johns Hopkins University Press.

Leto, V. (1988) '"Washing, Seems It's All We Do": Washing Technology and Women's Communication', in C. Kramarae (ed.) *Technology and Women's Voices*. London: Routledge & Kegan Paul.

Ling, R. (2004) *The Mobile Connection. The Cell Phone's Impact on Society*. San Francisco: Morgan Kaufman, Elsevier.

Lury, C. (1996) *Consumer Culture*. Cambridge: Polity.

Lutz, H. (2007) 'Editorial', *European Journal of Women's Studies* 14(3), 187–92.

Luxton, M. (1980) *More Than a Labour of Love*. Toronto, Ontario: The Women's Press.

LWC (London Women's Clinic) (2009) http://www.lwclinic.co.uk/gen/lesbian_single.php, accessed 30 June 2009.

Lykke, N. and Braidotti, R. (eds) (1996) *Between Monsters, Goddesses and Cyborgs. Feminist Confrontations with Science, Medicine and Cyberspace*. London: Zed Books.

Mackenzie, D. and Wajcman, J. (eds) (1985/1999) *The Social Shaping of Technology*. Milton Keynes: Open University Press.

Malcolmson, P. (1986) *English Laundresses. A Social History*. Urbana: University of Illinois Press.

Marcuse, H. (1964) *One Dimensional Man: Studies in the Ideology of Advanced Industrial Society*. Boston: Beacon Press.

Martens, L. and Scott, S. (2005) '"The Unbearable Lightness of Cleaning": Representations of Domestic Practice and Products in *Good Housekeeping* Magazine (UK): 1951–2001)', *Consumption, Markets and Culture* 8(4), 379–401.

Massey, D. (1992) 'A Place Called Home', *New Formations* 7, 3–15.

McCracken, E. (1993) *Decoding Women's Magazines*. London: Macmillan.

McDowell, L. (2008) 'The New Economy, Class Condescension and Caring Labour: Changing Formations of Class and Gender', *Nordic Journal of Feminist and Gender Research* 16(3), 150–65.

McDowell, L., Batnitzky, A. and S. Dyer (2007) 'Division, Segmentation and Interpellation: The Embodied Labours of Migrant Workers in a Greater London Hotel', *Economic Geography* 81(1), 1–26.

McNay, L. (1994) *Foucault: A Critical Introduction*. Cambridge: Polity.

McNay, L. (1999) 'Gender, Habitus and the Field: Pierre Bourdieu and the Limits of Reflexivity', *Theory, Culture and Society* 16(1), 95–117.

McNay, L. (2000) *Gender and Agency: Reconfiguring the Subject in Feminist and Social Theory*. Cambridge: Polity.

Miller, D. (1997), 'Consumption and its Consequences', in H. Mackay (ed.) *Consumption and Everyday Life*. London: Sage.

Miller, D. (ed.) (1998) *Material Cultures. Why Some Things Matter*. London: University College Press.

Miller, D. (ed.) (2001) *Home Possessions. Material Culture Behind Closed Doors*. Oxford: Berg.

Miller, D. and Slater, D. (2000) *The Internet. An Ethnographic Approach*. Oxford: Berg.

Mintel (1995) 'Laundry and Diswasher Appliances'. Market Intelligence, June, London: Mintel.

Mohun, A. (1996) 'Why Mrs. Harrison Never Learned to Iron: Gender, Skill and Mechanization in the American Steam Laundry Industry', *Gender and History* 8(2), 231–51.

Mohun, A. (1999) *Steam Laundries. Gender, Technology, and Work in the United States and Great Britain, 1880s-1940.* Baltimore: Johns Hopkins University Press.

Morgan, D. (1996) *Family Connections. An Introduction to Family Studies.* Cambridge: Polity.

Morgan, D. (1999) 'Risk and Family Practices, Accounting for Change and Fluidity in Family Life', in E.B. Silva and C. Smart (eds) *The 'New' Family?* London: Sage.

Morgan, D. (2004) 'Everyday Life and Family Practices', in E.B. Silva and T. Bennett (eds) *Contemporary Culture and Everyday Life.* Durham: Sociologypress, pp. 37–51.

Moore, N. (2007) '(Re)Using Qualitative Data?', *Sociological Research Online* 12(3), published 30 May 2007, http://www.socresonline.org.uk/12/3/1.html, accessed 19 August 2009.

Mulkay, M. (1997) *The Embryo Research Debate: The Science and Politics of Reproduction.* Cambridge University Press.

Murie, A. (1983) *Housing Inequality and Deprivation.* London: National Consumer Council.

Noble, D.F. (1977) *America by Design: Science, Technology, and the Rise of Corporate Capitalism.* New York: Knopf.

OM (2004) *Observer Magazine,* 'More Pain, More Gain', 18 April.

Ormrod, S. (1994) '"Let's Nuke the Dinner": Discursive Practices of Gender in the Creation of a New Cooking Process', in C. Cockburn and R. Fürst-Dilic (eds) *Bringing Technology Home: Gender and Technology in a Changing Europe.* Buckingham: Open University Press.

Osborne, T. and Rose, N. (1999) 'Do the Social Sciences Create Phenomena? The Example of Public Opinion Research', *British Journal of Sociology* 50(3), 367–96.

O'Toole, L. (1998) *Pornocopia: Porn, Sex, Technology and Desire.* London: Serpent's Tail.

Panasonic (1990) *Dimension 4. Cookery Book* (800W series). Copyrighted by Matsushita Electric Industrial Co. Ltd. Printed in Japan.

Parr, J. (1999) *Domestic Goods. The Material, the Moral, and the Economic in the Postwar Years.* University of Toronto Press.

PEP (Political and Economic Planning) (1945) *The Market for Household Appliances.* London: PEP and Oxford University Press.

Phillips, A. (1996) *Monogamy.* London: Faber and Faber.

Phillips, D.J. (2002) 'Negotiating the Digital Closet', *Information, Communication and Society* 5(3), 406–24.

Pillsbury, R. (1998) *No Foreign Food.* Boulder: Westview Press.

Plummer, K. (2008) 'Studying Sexualities for a Better World? Ten Years of Sexualities', *Sexualities* 11(1/2), 7–22.

Radiation Ltd. (1927) *Radiation Cookery Book. For Use with the Regulo New World Gas Cookers.* Birmingham: Radiation House.

Randall, S. (2002*) Resource Guide in Food and Society*, Learning and Teaching Support Network, http://www.heacademy.ac.uk/assets/hlst/documents/resource_guides/food_and_society.pdf, accessed 28 June 2009.

Ravetz, A. (with Turkington, R.) (1995) *The Place of Home. English Domestic Environments, 1914–2000.* London: E&FN Spoon.

Reay, D. (2000) 'A Useful Extension of Bourdieu's Conceptual Framework? Emotional Capital as a Way of Understanding Mothers' Involvement in their Children's Education', *The Sociological Review* 48(4), 568–85.

Richardson, D. (ed.) (1996) *Theorising Heterosexuality. Telling it Straight.* Buckingham: Open University Press.

Roberts, M. (1991) *Living in a Man-Made World. Gender Assumptions in Modern Housing Design.* London: Routledge.

Roseneil, S. and Mann, K. (1996) 'Unpalatable Choices and Inadequate Families: Lone Mothers and the Underclass Debate', in E.B. Silva (ed.) *Good Enough Mothering? Feminist Perspectives on Lone Motherhood.* London: Routledge.

Rybczynski, W. (1986) *Home – A Short History of an Idea.* Ontario: Penguin Books.

Savage, M., Bagnall, G. and Longhurst, B. (2004) 'Place, Belonging and Identity: Globalisation and the "Northern Middle Class"', in E.B. Silva and T. Bennett (eds) *Contemporary Culture and Everyday Life.* Durham: Sociologypress.

Sayer, A. (2005) *The Moral Significance of Class.* Cambridge University Press.

Sayer, A. (2010) 'Bourdieu, Ethics and Practice' Bourdieu' in E.B. Silva and A. Warde (eds) *Cultural Analysis and Bourdieu's Legacy: Settling Accounts and Developing Alternatives.* London: Routledge.

Schinkel, W. (2007) 'Sociological Discourse of the Relational: The Cases of Bourdieu and Latour', *The Sociological Review* 55(4), 707–29.

Schor, J. (1993) *The Overworked American.* New York: Basic Books.

Scott, J. (2008) *Women and Employment; Changing Lives and New Challenges.* London: Edward Elgar Publishing.

Scott, J.W. (1988) *Gender and the Politics of History.* New York: Columbia University Press.

Scranton, P. (1995) 'Determinism and Indeterminacy in the History of Technology', in M.R. Smith and L. Marx (eds) *Does Technology Drive History?* Cambridge, Mass: MIT Press.

Sevenhuijsen, S. (1998) *Citizenship and the Ethics of Care: Feminist Considerations of Justice, Morality and Politics.* London: Routledge.

Short, F. (2006) *Kitchen Secrets. The Meaning of Cooking in Everyday Life.* Oxford: Berg.

Shove, E. (2003) *Comfort, Cleanliness and Convenience. The Social Organization of Normality.* Oxford: Berg.

Shove, E. and Southerton, D. (2000) 'Defrosting the Freezer: From Novelty to Convenience', *Journal of Material Culture* 5(3), 301–19.

Silva, E.B. (1988) 'Labour and Technology in the Car Industry. Ford Strategies in Britain and Brazil', PhD dissertation, University of London.

Silva, E.B. (1991) *Refazendo a Fábrica Fordista. Contrastes da indústria automobilística no Brasil e na Grã-Bretanha* (Remaking the Fordist Factory. Contrasts between the Car Industry in Brazil and Britain). São Paulo: Hucitec/Fapesp.

216 *Bibliography*

Silva, E.B. (1997a) '"This is the Way We Wash..." Laundry and Dish Washing in Britain, 1900-1990s', *GAPU Research Working Paper* no. 17, School of Sociology and Social Policy, University of Leeds.

Silva, E.B. (1997b) 'Household Technologies: Patterns of Innovation and Gender Relations', ESRC End of Award Report, R000231700, 31 March.

Silva, E.B. (1999a) 'Co-Applicant Final Report' to the ESRC, Project 'A Virtual Ethnography of the Dynamics of Social Change in Relation to New Technology', L132251048, September, unpublished.

Silva, E.B (1999b) 'Transforming Housewifery: Dispositions, Practices and Technologies', in E.B. Silva and C. Smart (eds) *The 'New' Family?* London: Sage, pp. 46-65.

Silva, E.B. (2000a) 'The Cook, the Cooker and the Gendering of the Kitchen', *The Sociological Review* 48(4), 612-28.

Silva, E.B (2000b) 'The Politics of Consumption @ Home', *Pavis Papers* no. 1. Open University.

Silva, E.B. (2002a) 'Time and Emotion in Studies of Household Technologies', *Work, Employment and Society* 16(2), 329-40.

Silva, E.B. (2002b) 'Routine Matters: Narratives of Everyday Life in Families', in G. Crow and S. Heath (eds) *Social Conceptions of Time. Structure and Process in Everyday Life*. Basingstoke: Palgrave Macmillan, pp. 179-94.

Silva, E.B. (2004) 'Materials and Morals: Families and Technologies in Everyday Life', in E.B. Silva and T. Bennett (eds) *Contemporary Culture and Everyday Life*. Durham: Sociologypress, pp. 52-70.

Silva, E.B. (2005) 'Gender, Home and Family in Cultural Capital Theory', *British Journal of Sociology* 56(1), 83-103.

Silva, E.B. (2006) 'Homologies of Social Space and Elective Affinities: Researching Cultural Capital', *Sociology* 40(6), 1171-89.

Silva, E.B. (2007a) 'Gender, Class, Emotional Capital and Consumption in Family Life', in Casey, E. and Martens, L. (eds) *Gender and Consumption: Domestic Cultures and the Commercialisation of Everyday Life*. Aldershot: Ashgate, pp. 141-62.

Silva, E.B. (2007b) 'What's [Yet] to Be Seen? Re-Using Qualitative Data', *Sociological Research Online* 12(3), published 30 May 2007, http://www.socresonline.org.uk/12/3/4.html, accessed 19 August 2009.

Silva, E.B, Warde, A. and Wright, D. (2009) 'Using Mixed Methods for Analyzing Culture: The Cultural Capital and Social Exclusion Project', *Cultural Sociology* 3(2), 299-316.

Silva, E.B. and Wright, D. (2005) 'The Judgement of Taste and Social Position in Focus Group Research', *Sociologia e Ricerca Sociale,* special double issue, 76/77, pp. 241-53. Milan: Angeli. Also available at: http://www.open.ac.uk/socialsciences/includes/__cms/download.php?file=w4lkl1zo6ef8h2c0b0.pdf&name=judgement_of_taste_and_social_position_in_focus_group_research.pdf, accessed 19 August 2009.

Silva, E.B. and Wright, D. (2009) 'Display, Desire and Distinction in Housing', *Cultural Sociology* 3(1), 31-50.

Silver, H. (1987) 'Only So Many Hours in a Day: Time Constraints, Labour Pool and Demand for Consumer Services', *Service Industries Journal* 7(4), 26-45.

Silverstone, R., Hirsch, E. and Morley, D. (1992) 'Information and Communication Technologies and the Moral Economy of the Household', in R. Silverstone and E. Hirsch (eds) *Consuming Technologies*. London: Routledge.

Simon, A. and Whiting, E. (2007) 'Using the FRS to Examine Employment Trends of Couples', *Economic and Labour Market Review* 1(11), http://www.statistics.gov.uk/elmr/11_07/downloads/ELMR_Nov07_Simon.pdf, accessed 16 March 2009.

Simpson, M. (1994) 'A World of Penises', in *Male Impersonators: Men Performing Masculinity*. London: Cassell.

Slater, D. (1997) *Consumer Culture and Modernity*. Cambridge: Polity.

Smart, C. (1996) 'Deconstructing Motherhood' in E.B. Silva (ed.) *Good Enough Mothering? Feminist Perspectives on Lone Motherhood*. London: Routledge.

Smart, C. (2007) *Personal Life*. Cambridge: Polity.

Smeds, R., Huida, O., Haavio-Mannila, E. and Kauppinen-Toropainen, K. (1994) 'Sweeping Away the Dust of Tradition: Vacuum Cleaning as a Site of Technical and Social Innovation', in C. Cockburn and R. Fürst-Dilic (eds) *Bringing Technology Home: Gender and Technology in a Changing Europe*. Buckingham: Open University Press.

Smith, D. (1987) *The Everyday World as Problematic: A Feminist Sociology*. Milton Keynes: Open University Press.

Social Focus on Women and Men (1998) Central Statistical Office, London: HMSO.

Social Trends (2008) no. 38, Central Statistical Office, London: HMSO.

Southerton, D. (2001) 'Consuming Kitchens. Taste, Context and Identity Formation', *Journal of Consumption Culture* 1(2), 179–203.

Spigel, L. (2005) 'Designing the Smart House. Posthuman Domesticity and Conspicuous Production', *European Journal of Cultural Studies* 8(4), 403–26.

SRoL (2007) *Sociological Research Online*, 12(3), Refereed Special Section: 'Reusing Qualitative Data', http://www.socresonline.org.uk/12/3/.html, accessed 19 August 2009.

Stanley, L. (1995) 'Women Have Servants and Men Never Eat: Issues in Reading Gender, using the Case Study of Mass-Observation's 1937 Diaries', *Women's History Review* 4(1), 85–102.

Star, S.L. (1991) 'Power, Technology and the Phenomenology of Conventions: On Being Allergic to Onions', in J. Law (ed.) *A Sociology of Monsters: Essays on Power, Technology and Domination*. London: Routledge.

Strasser, S. (1982) *Never Done: A History of American Housework*. New York: Random House.

Strasser, S., McGovern, C. and Judt, M. (1998) 'Introduction', in S. Strasser, C. McGovern and M. Judt (eds) *Getting and Spending. European and American Consumer Societies in the Twentieth Century*. Cambridge University Press, pp. 1–8.

Strathern, M. (1999 [1992]) 'Introduction. First edition: A Question of Context', in J. Edwards, S. Franklin, E. Hirsch, F. Price and M. Strathern, *Technologies of Procreation. Kinship in the Age of Assisted Conception*. London: Routledge, pp. 9–28.

Sullivan, O. (2000) 'The Division of Domestic Labour: 20 Years of Change?', *Sociology* 34(3), 437–56.

Sutherland, D.E. (1981) *American and their Servants. Domestic Service in the United States from 1800 to 1920*. Baton Rouge: Louisiana State University Press.

Swasy, A. (1993) *Soap Opera. The Inside Story of Procter and Gamble*. New York: Simon and Schuster.

Swinbank, V.A. (2002) 'The Sexual Politics of Cooking: A Feminist Analysis of Culinary Hierarchy in Western Culture', *Journal of Historical Sociology* 15(4), 464–94.

Terry, J. and Calvert, M. (eds) (1997) *Processed Lives. Gender and Technology in Everyday Life*. New York: Routledge.

The Guardian (newspaper) – various years and numbers.

The Lady (magazine) – various years and numbers.

The Observer (newspaper) – various years and numbers.

Thévenot, L. (2001) 'Pragmatic Regimes Governing the Engagement with the World', in T.R. Schatzki, K. Knorr Cetina and E. von Savigny (eds) *The Practice Turn in Contemporary Theory*. London: Routledge.

Thévenot, L. (2007) 'The Plurality of Cognitive Formats and Engagements: Moving Between the Familiar and the Public', *European Journal of Social Theory* 10(3), 409–23.

Thévenot, L. and Boltanski, L. (2006) *On Justification. Economies of Worth*. Princeton University Press.

Thrall, C.A. (1970) 'Household Technology and the Division of Labor in Families', PhD Thesis, Harvard University.

Tronto, J.C. (1993) *Moral Boundaries. A Political Argument for an Ethics of Care*. New York: Routledge.

Turkle, S. (1995) *Life on the Screen: Identity on the Age of the Internet*. New York: Simon and Schuster.

Urry, J. (2000) 'Mobile Sociology', *British Journal of Sociology* 51(1), 185–203.

Urry, J. (2004) 'Connections', *Environment and Planning D: Society & Space* 22, 27–37.

Vanek, J. (1973) 'Keeping Busy: Time Spent on Housework, United States, 1920–1970', unpublished PhD Thesis, University of Michigan.

Wajcman, J. (1991) *Feminism Confronts Technology*. Cambridge: Polity.

Wajcman, J. (2002) 'Addressing Technological Change: The Challenge to Social Theory', *Current Sociology* 50(3), 347–63.

Walkerdine, Valerie (1997) *Daddy's Girl*. London: Macmillan.

Warde, A. (1990) 'Household Work Strategies and Forms of Labour: Conceptual and Empirical Issues', *Work, Employment and Society* 4(4), 495–515.

Warde, A. (1994) 'Consumption, Identity – Formation and Uncertainty', *Sociology* 28(4), 877–98.

Warde, A. (1997) *Consumption, Food and Taste*. London: Sage.

Warde, A. and Martens, L. (2000) *Eating Out. Social Differentiation, Consumption and Pleasure*. Cambridge University Press.

West C. and Zimmerman, D. (1987) 'Doing Gender', *Gender & Society* 1, 125–51.

Wetherell, M. (2006) 'Formulating Selves: Social Psychology and the Study of Identity', *Social Psychological Review* 8(2), 62–73.

What Price? (2009) 'Nappy Costs', http://www.whatprice.co.uk/health/parent/nappies.html, accessed 3 April 2009.

Which? (magazine) – various years and numbers.

Williams, L. (1989) *Hard Core: Power, Pleasure and the 'Frenzy of the Visible'*. London: Pandora.

Williams, R. (1977) *Marxism and Literature*. London: Oxford University Press.

Winner, L. (1988 [1980]) 'Do Artefacts Have Politics?', in L. Winner (ed.) *The Whale and the Reactor: A Search for Limits in an Age of High Technology*. University of Chicago Press, pp. 19–39.

Winnicott, D.W. (1963) 'Communicating and Not Communicating Leading to a Study of Certain Opposites', in *The Maturational Processes and the Facilitating Environment*. London: Hogarth Press.

Wolfreys, J. (2000) 'In Perspective: Pierre Bourdieu', *International Socialism Journal* 87, http://pubs.socialistreviewindex.org.uk/isj87/wolfreys.htm, accessed 15 December 2009.

Woman's Own (magazine) – various years and numbers.

Wood, R. (ed.) (2000) *Strategic Questions in Food and Beverage Management;* London: Butterworth-Heinemann.

Young, M. and Willmott, P. (1973) *The Symmetrical Family*, Harmondsworth: Penguin.

Zukin, S. (1990) 'Socio-spatial Prototypes of a New Organisation of Consumption: The Role of Real Cultural Capital', *Sociology* 24, 37–56.

Index